EXTINCTION DIALOGS

EXTINCTION DIALOGS

HOW TO LIVE WITH DEATH IN MIND

CAROLYN BAKER & GUY McPHERSON

NEXT REVELATION PRESS

AN IMPRINT OF

TAYEN LANE

PUBLISHING

San Francisco, Montreal

www.tayenlane.com

COPYRIGHT

Edited by Bri Bruce
Cover Design: Tayen Lane Design
Cover photographs courtesy Shutterstock

DEDICATION

To the memory of our friend, Michael C. Ruppert,
and the shards of the living planet we leave in our wake.

TABLE OF CONTENTS

EXTINCTION DIALOGS

HOW TO LIVE WITH DEATH IN MIND

FOREWORD

No one who truly reads this book will ever be the same afterward. Whether you agree in the end with its drastic and shocking conclusions is, of course, up to you; the very fact that such a book outlining the case for the coming extinction of life on Earth could be written at all—and with such fearless clarity and compassion—changes absolutely all serious conversation about our contemporary world crisis and our potential response to it.

In *Extinction Dialogs*, Carolyn Baker and Guy McPherson make four essential points. The first is that overwhelmingly scientific evidence now makes it clear that our environment is headed for swift apocalyptic collapse, and that this evidence now has to be faced and integrated into the deepest level of the psyche by all those human beings brave enough who want to face the unspeakable facts rather than be drugged by what Guy McPherson wryly calls "hopium." The authors know from their own fierce experience that such integration is extremely costly and demands a capacity to weather unprecedented grief, depression, outrage, and despair. The authors believe there is no way out of this extreme process; we can only be of help to each other in the end game of our world if we allow all of our remaining illusions about human agency, fix-it solutions, and magical thinking about divine intervention to be ruthlessly burned away in the furnace of truth.

The second theme this book explores fearlessly is that as we do face the facts about climate change; we find ourselves staring into the face of death, not only of our civilization, but of our species and all animal species and most of nature. What Guy and Carolyn make clear, however, is that this facing of the unimaginable and unspeakable need not, and must not, result in despair, helplessness, and destructive passivity. Beyond hope lies love—love for its own sake and truth and radiance, love that we must learn to embody and act out on every level in every way to find meaning and joy in the final situation we have created from our ignorance and greed. What the now likely scenario of extinction is demanding of all of us is going beyond our human fantasies of entitlement, success, and survival into a radical embodiment of what I would call the divine truth of our nondual identity—a truth that will be all the more inspiring and empowering in a situation when it alone can provide strength, calm, and continuing purpose. Millions of us will now have to urgently undertake the journey of awakening for the world not to die out in a brutal orgy of madness and chaos.

The third theme that Guy and Carolyn drive home is that we must now find the courage to accept that the whole world could very well be in a "hospice" situation already and that this asks all of us to make the most serious decision now: how to live the rest of our lives in love and service, not for the sake of success or survival but to keep human dignity and compassion alive in extreme circumstances. We, all of us, must open our hearts in all-embracing compassion toward all the other beings trapped in what Rilke called "this burning boat of meat," practicing active kindness toward all the animals that will be incinerated in an apocalypse of our own making, and doing everything in our own power to reach out to the weaker among us and those that will be driven to despair, and even madness, as the full truth of our appalling predicament becomes diamond-clear. Sacred activism becomes even more important and soul saving in a terminal situation for two reasons. The first is that action is always the best antidote to despair, even when it cannot change the circumstances. Secondly, action from a sacred consciousness is the one way in which, even if we have to die out, we can do so with dignity and without forfeiting the one thing we can still preserve: the essential beauty of our divine human truth. The lesson that Victor Frankl learned in the horror of the concentration camps—that only love in action can provide strength and meaning in hopeless disaster—will now have to be learned by all of us if our species is not to die out in horrible chaos and violence. One of the most moving and inspiring aspects of *Extinction Dialogs* is how Guy and Carolyn, again and again, show us the way to finding joy and living compassion for their own sake, and so empowering ourselves with the strength that can be born in us if we are fearless enough to face where we are and what can and still must be done.

The fourth theme that dances like a golden thread throughout this book is that when we do accept our potentially terminal fate an extreme love for life on its own terms—and just as it is—can be exploded within us, along with a radiant gratitude for the simplest things we have taken for granted and a rapture at the beauty of the world we are losing. When we finally face that time is running out, not just for the human race but for all life, we can—if we choose and if we pray and meditate deeply and continue to act humbly and with unconditional love in whatever circumstances that we find ourselves in—live in peace and joy and surrender to mystery. As I myself have opened to more and more of what Carolyn and Guy are trying to tell us, I have found that it has made me want to serve more wisely, give more selflessly, and live more consciously in the calm and joyful depths of myself that death and defeat cannot pervert or destroy; all my life I have worked passionately to try and prevent the very disaster that is now upon us, and I have found that finally accepting the possibility of extinction has not destroyed my inner resolve but purified, matured, and honed it in ways that I could never have foreseen. I have come to know, from my inmost experience, that we are never more deeply accompanied by divine grace than in extreme circumstances and that while divine grace may not save us from facing the consequences of our own terrible choices and ignorance, it can do something even more miraculous: lift us free of our false selves, initiate us into the true, deep self that loves and gives and serves anyway, come what may. This, I am discovering slowly and sometimes very painfully, is liberation.

Please read this book several times carefully and with compassion for yourself and great gratitude for the two humbly heroic human beings that here have joined their voices to proclaim a terrifying but potentially all-transforming truth. And when you have read this book and have started to integrate its message, give copies to everyone you know and start to accompany them in their journey to opening to and enacting with grace its ferocious but liberating reality.

- Andrew Harvey, Author of Radical Passion
North Atlantic Books, 2012

INTRODUCTION

Civilization is not the only way to live. Indeed, humans lived without civilization for more than two million years. We've lived within the shackles of civilization for a few thousand years. Civilization clearly is omnicidal. Few notice. Even fewer care.

Civilization is an expression of patriarchy. The current version of industrial civilization benefits a few Caucasian men at the expense of every other living being. Most civilized people believe this set of living arrangements is wonderful.

If you are reading these words, you benefit from imperialism. American Empire is real, and it covers the globe. There is no escape.

Born into captivity and assimilated into the normal bias of a historically abnormal period in world history, for many years we did all the things this culture expected from us. For example, we began our careers in the expected manner, as classroom conservatives. We received accolades and numerous awards for teaching, advising, and scholarship. Early on we realized students don't care what you know until they know you care— about them. And we did in ways that made our colleagues question whose side we were on even while we were pointing out that, in educating ourselves and others, we're all on the same side.

After a few years as classroom conservatives, we learned something important: even earnest, caring teaching doesn't necessarily lead to learning. The Sage on the Stage approach is dead. So, too, is the model of student as customer. So we each switched to a model based on a "Corps of Discovery" in which every participant is expected to contribute to the learning of every other participant. We practiced anarchism, in our own classroom-centered way, taking responsibility for ourselves and our neighbors. This radical approach to teaching puts it all on the line. Everything we know and everything we are is exposed during every meeting of every class. How can we evaluate our knowledge, our wisdom, and our own personal growth without exposing our assumptions at every turn? This, of course, requires us to let go—to let go of our hubris and replace it with humility, to let go of our egos and instead seek compassion, perhaps even empathy.

We're all on the same side. We are the ones we've been waiting for, as the saying goes, but we must let go of a system that is making us sick, making us crazy, and is killing us. During our entire lives, success has been defined, incorrectly, by the amount of money rather than the amount of compassion a person has. Similarly, the entire system has been defined in terms that make no sense because the system rewards money over joy, and death over life. As John Ralston Saul pointed out in his 1992 book *Voltaire's Bastards,* "never has failure been so ardently defended as success (Saul, 1992)."

Fortunately, we're headed to a world where money doesn't matter. And without money we'll all be rich in the life-sustaining ways that really matter. We spent the first couple million years of the human experience immersed in gift economy, and it seems we'll be there again in the not-too-distant future. We long for the day we see more free-flowing rivers every year, as well as more clean air, more wild places, fewer species driven to extinction, and less soil washed into the world's oceans.

Like us, everybody in the industrialized world was born into captivity. A few people seem to be acknowledging the bars imprisoning them. The unseen bars of our prisons are keeping us from becoming fully human, from fully expressing our humanity. As Goethe said some two centuries ago, "None are more hopelessly enslaved than those who falsely believe they are free (Goethe, 1809)." It's time for a jailbreak. Better yet, it's time to heed the words of Joan Baez and raze, raze the prisons to the ground, especially the invisible prisons.

More than a decade ago our scholarly efforts had broadened to include the twin sides of the fossil-fuel coin—global climate change and peak oil—and our messages increasingly targeted the public. Long-time teachers, we had become friends of the Earth as well as social critics. And we'd come to recognize the costs and consequences of the industrial economy: obedience at home, oppression abroad, and wholesale destruction of the living planet on which we depend for our very lives. We're on track to cause our own extinction, probably within a matter of two decades because of ongoing climate change. The only legitimate hope to prevent our near-term extinction, and that of the thousands of species we're taking with us into the abyss, is completion of the ongoing collapse of the industrial economy. Even that may not suffice, and it will trigger the catastrophic meltdown of the world's nuclear facilities, which number more than four hundred.

The omnicidal culture we know as Western civilization is about to reach its overdue end. Time is not on our side. It's long past time to let go of a system that enslaves us all while destroying all life and therefore all that matters. And it's not merely time to let go, but to terminate this increasingly violent system that values the property of the rich more than the lives of the poor.

As Arundhati Roy wrote in her 2001 book, *Power Politics,* "Trouble is that once you see it, you can't unsee it. And once you've seen it, keeping quiet, saying nothing, becomes as political an act as speaking out. There's no innocence. Either way, you're accountable. (Roy, 2001)"

We put our own spin on Arundhati Roy's conclusion: "Big Energy poisons our water. Big Ag controls our seeds, hence our food. Big Pharm controls, through pharmaceuticals, the behavior of our children. Wall Street controls the flow of money. Big Ad controls the messages we receive every day. The criminally rich get richer through crime: that's how America works. Through it all, we believe we're free (Roy, 2001)."

From, *Anam Cara: A Book of Celtic Wisdom*, John O'Donohue writes:

> *There is a presence who walks the road of life with you. This presence accompanies your every moment.*

It shadows your every thought and feeling. On your own or with others, it is always there with you. When you were born, it came out of the womb with you, but with the excitement at your arrival, nobody noticed it. Though this presence surrounds you, you may still be blind to its companionship. The name of this presence is death. (O'Donohue, 1998)

John O'Donohue knew only too well the truth of these sentences. He left us in 2008 at the age of fifty-two. The Irish poet, scholar, and author was a child of the indigenous Celtic tradition, which like so many other Earth-based traditions grasps on a cellular level that death is a part of life. In the same tradition, storyteller and mythologist Michael Meade often states in his public lectures that "Life is fatal; no one gets out of here alive."

For several years, people have sought us out for life coaching. Generally, these are folks who have done extensive research on the collapse of industrial civilization and near-term extinction. Often they are surrounded by other individuals who cannot bear to hear what they have learned about collapse, and they want to speak with us in order to validate their understanding of collapse and extinction and to find others, even if they are in another country, with whom they can share their thoughts and feelings. We, too, have a small circle of friends with whom we can discuss these issues. As for Boulder, Colorado, where Carolyn lives, one might believe otherwise given that it is an environmentally conscious, purportedly progressive community. Yet dinner conversation about the end of our industrially civilized living arrangements—or worse, the extinction of species, including our own—is certain to produce indigestion and exile from even the most radical circles in town. In rural Southern New Mexico, where Guy lives, the human community embraces any topic of conversation. But, as with any bitter pill, a little near-term human extinction goes a long way.

We are often asked the following question: Why do some people wake up and others do not? We have no answer to that except to say that those who refuse to examine the realities screaming in their faces are, among other things, terrified of death. We will say more about that in subsequent pages, but what we know with certainty is that when people can contemplate their own death and that of their loved ones and lifestyle, waking up is not only easier but often obligatory.

For most members of this culture, literal death is some far-off exigency that is only a concept in the mind and one that they hope to keep as much distance from as possible. Yet in a death-phobic culture, individuals do not recognize symbolic deaths as an extension of "the big one." Of this, John O'Donohue writes in *Anam Cara*, "In a certain sense, the meeting with your own death in the daily forms of failure, pathos, negativity, fear, or destructiveness are actually opportunities to transfigure your ego."(O'Donohue, 1998)

I will be commenting much more on what Joanna Macy calls "the greening of the ego" in the following pages, but we want to clarify at the outset that the ego is not a bad thing. We all need one. Try finding the car keys or the bathroom without one. However, most indigenous traditions teach their young from very early on that they came here with phenomenal gifts that the community needs but which cannot be expressed or utilized unless everyone in the community is committed to engaging with and nurturing the deeper self. In order for this to occur, the ego must be tempered—gradually learning to occupy one of the seats in "coach class" as opposed to the cockpit.

Today we stand on the threshold of our own literal death and the death of nearly every living being on the planet. We want you to read this book and live with death in mind, not just because we occupy this grotesque place in human history, but because of what living with death in mind can do for all of us. Again, John O'Donohue states (O'Donohue, 1998):

> *Death is the great wound in the universe and the great wound in each life. Yet, ironically, this is the very wound that can lead to new spiritual growth. Thinking of your death can help you to radically alter your fixed and habitual perception. Instead of living according to the merely visible material realm of life, you begin to refine your sensibilities and become aware of the treasures that are hidden in the invisible side of your life. (O'Donohue, 1998)*

Like most contemporary Americans, we believed we were free far too long, but we were in fact bound by the monkey trap. A monkey trap is a small cage with a piece of fruit inside, anchored to a solid object. The cage has a hole barely large enough for a monkey to insert its empty hand, but too small to extract the hand holding a piece of fruit. The monkey is trapped, unable to let go of the fruit.

It's time to let go of the low-hanging fruit of empire. It's time to live as human animals, fully immersed in lives of excellence.

- Guy R. McPherson, from the Mud Hut in Southern New Mexico

- Carolyn Baker, from the foothills of the Colorado Rocky Mountains

CHAPTER 1

FROM PROFESSORS TO PROPHETS

*The only book that is worth writing is the one we
don't have the courage or strength to write. The book
that hurts us (we who are writing), that makes us
tremble, redden, bleed.*

~Helen Cioxous~

CAROLYN'S JOURNEY

On a sunny fall afternoon in October 2000, I sat at my
computer in El Paso, Texas Googling "CIA and drugs." The
reason for my search has long since escaped my memory, but
ultimately my quest led me to connect with Mike Ruppert. At the
time, Mike was living in Los Angeles and publishing his *From
The Wilderness* newsletter to which I soon subscribed. In
December 2000, I met Mike and a fourteen-year professional
connection ensued.

As the reader may recall, this was the decade of the
George W. Bush presidential "selection," September 11, the Iraq
War, the beginning of peak oil awareness, the housing bubble,
and the financial collapse of 2008. Throughout the decade, as an
adjunct college history professor I was riveted by the historical
implications of these events. In 2006, I published a supplemental
textbook titled *US History Uncensored: What Your High School
Textbook Didn't Tell You.* Yet it was not until 2007 that I ceased

viewing the myriad issues of the time as separate from each other, but rather as part of a web of circumstances that spelled the collapse of industrial civilization. My awareness was expedited by Tim Bennett and Sally Erickson's documentary What a Way to Go: Life at the End of Empire. After viewing the film numerous times, collapse was crystalized in my consciousness.

Before becoming a history and psychology professor, I had been a psychotherapist for seventeen years. My training in Jungian depth psychology deeply influenced my perception of collapse. Once I grasped its reality, I immediately began contemplating the emotional and spiritual implications for the average inhabitant of industrial civilization. What would be the impact on that person? Collapse would mean catastrophic loss of life and societal disruption—the termination of life as most modern humans had known it. What emotions would be evoked? In the midst of chaos, how might people find meaning and purpose—or would that even be possible?

I began making written notes that quickly turned into my first book on the topic, *Sacred Demise: Walking the Spiritual Path of Industrial Civilization's Collapse*. Given the content of the book I presumed that it would sell few copies and be wildly rejected even by those who understood collapse. On the contrary, the book was embraced by many more individuals than I had imagined, and I published *Navigating the Coming Chaos: A Handbook for Inner Transition in 2011*, which took *Sacred Demise* to the next level in a number of ways, not the least of which was the fact that Navigating the Coming Chaos provided a superb format for groups who wanted to meet and study collapse preparation together.

In 2007, Guy McPherson, who at the time was Professor of Natural Resources, Ecology, and Evolutionary Biology at the University of Arizona, contacted me. He told me he would be traveling to where I lived in Las Cruces, New Mexico to give a presentation on peak oil at a conference at New Mexico State University. Guy and I met for coffee and discussed our shared perspectives in relation to collapse. Meanwhile Guy continued his life as professor until his resignation from the University of Arizona a few years later and his establishment of an off-the-grid homestead in Western New Mexico. We did not reconnect until Guy was well into his compilation of climate research in 2013.

From 2007 to 2013, I focused on the collapse of industrial civilization. I knew that climate change was a huge piece of the equation, but it wasn't until the summer of 2012 with the mega-drought in the United States that I began to delve more deeply into climate science. In 2013, I encountered Guy again at a conference at which we were both presenting. While there I asked if I might visit his New Mexico homestead that he calls "The Mud Hut." A month later I briefly visited the Mud Hut, and subsequent to that visit Guy spent several days giving presentations in my hometown, Boulder, Colorado in October 2013. As my professional connection with Guy deepened, so did my exploration of his climate research. By the end of 2013, I was convinced that it is highly unlikely that most of the human species will survive climate catastrophe beyond 2050, with very few humans surviving until 2100.

Although I had the greatest respect for Guy on every level, I was not familiar with his spiritual perspective or even whether or not he had one. Whereas we had traded our own books with each other, and Guy told me that he had the greatest respect for my work, we never discussed his spirituality and I was not inclined to do so, wanting to avoid any hint of proselytizing him. While in Boulder, however, Guy announced to his allies that he had been invited to present in Winnipeg the following month and would be staying with a group of Buddhist monks there. Two months later in January 2014, Guy participated in a Grief Recovery Training Program and announced that he had become a certified grief counselor.

What matters most to me in relating what I know of Guy's encounter with the Buddhists in Winnipeg and the Grief Recovery Training is not the content of those experiences, some of which Guy shares in this book, but the changes I observed in Guy as a result of them. While I do not wish to name specifics, I witnessed in general a profoundly softened human being who appeared to me less bitter, burdened, and brittle in his awareness of near-term human extinction. He began more frequently using the word "acceptance" in his work, and I was beginning to understand why. Encountering the reality of extinction—his own along with the rest of humanity—had hastened and deepened the journey from head to heart.

In 2012, I was confronted with a recurrence of breast cancer and opted to have a double mastectomy. The cancer was caught very early and the tumor was very small, but facing cancer again after nineteen years of not having to do so was greatly life altering for me. I made radical changes in diet and exercise and perhaps for the first time in my life came to fully inhabit my body.

As my friend Michael Meade writes, "Life is fatal; no one gets out of here alive." The recurrence of cancer, particularly at the age of sixty-seven, forced me once again to confront my own mortality. It was a clarion wake-up call which compelled me to prioritize the work to which I feel called and the supportive relationships in my life that sustain me as I do it. It was the beginning of a captivating dance with death with which I intend to engage for the remainder of my days on this planet. As a result, when Guy or others to whom we may refer in this book speak of "Earth in hospice," I resonate instantly and effortlessly.

In May 2013, a close friend of mine in New Mexico passed away after spending nearly a month in hospice. She and I had had a falling out two years prior, but only a month before her death we had reconciled and shared one of the most intimate conversations we had ever had. I had the privilege of speaking with her every other day by phone during the last month of her life. On April 30, 2013, we spoke for the last time. Her last words to me were, "My heart is with you."

In the final month of my friend's life, I had the opportunity to witness from a distance her experience in hospice. She was surrounded with a host of supportive friends and superbly trained hospice nurses who gave her more nurturing than she had ever experienced in her sixty-six years on Earth. I utilized my friend's experience, alongside other accounts of individuals known and unknown to me who have related their hospice experiences prior to their passing, in order to begin writing more directly about the likelihood that all species, including humans, are now abiding in a hospice situation which we can choose to accept or resist. Much of that writing is incorporated into this book.

In April 2014, my friend and colleague Mike Ruppert, who was the first human being to introduce me to the reality of collapse, took his own life. I was fortunate to become part of Mike's inner circle of associates, and all of us in that circle were aware of Mike's obsession with suicide and the likelihood that someday he would act upon it. His death has shaken me to my core, as it has shaken countless individuals who followed his work and who are grappling with near-term extinction. Among the many issues it has forced all of us to confront is our mortality and our fragility as humans living in the death grip of industrial civilization and its pillage of our planet.

Obviously I have been researching the collapse of industrial civilization specifically since 2007, but I have in fact been, even without realizing it, exploring the reality of a planet in hospice for the past fourteen years. It was one thing to read and write about what James Howard Kunstler named "The Long Emergency," and quite another to grasp the likelihood that the "sacred demise" of which I wrote in 2009 would not be a long, slow tumble down a bumpy hill, but rather a much more acute catastrophe that actually spelled near-term human extinction. Not simply to "realize" but to be taken captive by the likelihood of near-term extinction has altered every aspect of my life. No activity, no relationship, no pastime in my life, however seemingly routine, escapes its influence. Whether or not I ever have another encounter with cancer, I am in fact living in hospice, as are all beings on Earth. Rather than evoking depression, malaise, suicidal ideation, or despair in me, my "voluntary admission to hospice" has brought more inspiration, aliveness, compassion, and joy than I could ever have imagined. I am now beginning to grasp the wisdom of the following African proverb: When death comes, may it find you fully alive.

GUY'S PATH

As I explained in my 2013 book, *Going Dark*, until recently I believed complete collapse of the world's industrial economy would prevent runaway greenhouse and therefore allow our species to persist for a few more generations. But in June 2012, the ocean of evidence on climate change overwhelmed me, and I no longer subscribe to the notion that habitat for humans will exist on Earth beyond the 2030s. We've triggered too many self-reinforcing feedback loops to prevent near-term extinction at our own hand, as I explain in a later chapter.

This was actually a *déjà vu* moment for me. I had reached a similar conclusion in the early days of the twenty-first century while editing a book about climate change. About a year after I realized we were "done" and I began mourning the impending loss of our species, I discovered what appeared to be the ultimate hail-Mary pass: the impending peak in global oil production might cause industrial civilization to collapse in time to prevent near-term human extinction. Alas, several years after passing the peak of global crude oil extraction in 2005 or 2006 (according to the US Energy Information Administration and International Energy Agency, respectively), the evidence is in: we're done all over again. The mourning began anew for me in June 2012.

About this same time, I realized that my carefully thought out and constructed living arrangements no longer made any sense. My project of abandoning empire and moving off grid to a shared property in New Mexico has failed catastrophically. Consider the primary reasons I left the easy life of a tenured professor at a major university to develop and occupy the property I call the Mud Hut:

1. an act of resistance against the dominant paradigm (the dominant paradigm and those within it failed to notice);

2. an example of alternative living, in my case promoting a gift economy within agrarian anarchy (my example has failed to inspire a significant number of others to live differently);

3. a way to provide more time for speaking and writing about important topics and actions that were discouraged at the university (I have enjoyed limited success in this

arena, although time freed up not battling administrative dragons has been largely consumed with rigorous physical work);

4. a refuge for the young son of the couple with whom my wife and I share this property, as well as his generation (due to ongoing, accelerating climate change the youngster's future in this location probably will be notably short), and

5. a way to extend my own life, and that of my wife (due to ongoing, accelerating climate change, our future in this location is likely to be quite short).

With respect to the latter two items on the list, it is important to note the relationship between collapse of industrial civilization and abrupt climate change. According to the analysis in Clive Hamilton's April 2013 book *Earthmasters*, the demise of industrial civilization will cause the planet to warm an additional 1.1 C within very little time. Adding the 1.1 C warming to the 0.85 C warming Earth has experienced since the beginning of the industrial revolution takes us to the much-dreaded political target of 2 C—which has never been a scientific target, although many mainstream climate scientists ignorantly proclaim as much—suggesting abrupt warming in excess of 5 C here in the southwestern portion of a large continent in the Northern Hemisphere. I cannot imagine human life surviving such an event beyond a few months.

Perhaps most importantly there is a widening chasm within the partnership formed on this shared property. Although leaving the life I loved as an academic to move to this location made sense at the time, long before the climate change news grew so dire and when collapse of industrial civilization appeared imminent, I now view the major personal transition as infinitely regrettable. I have come to see it as the worst mistake of my life so far. I'm not dead, though, so I suspect I can yet outdo myself.

Looking back on my enormous effort and life-changing transition, it seems that perhaps our choice of homesteading partners was doomed to fail. Partnerships of all sorts fail more often than not (consider, for example, the divorce rate in the United States). The choice of partners for constructing a viable off-grid homestead is perhaps even more crucial than the choice of one's spouse.

My attempt to walk away from empire has failed in part because empire is stunningly difficult to leave behind. The disaster of American imperialism follows like the "evil twin" of a beloved dog: you know something is dogging your heels, but it's no longer your loving companion. I'm left with the horrors of American Empire and the abyss of near-term human extinction without man's best friend.

I'm back to teaching. I'm not a teacher because it allows me to make money or because it gives me something to do. Indeed, I earn no money from my teaching, and I have plenty of tasks on the homestead. I'm a teacher because it's what I am, not what I do. Teaching is within me. It cannot be separated from my inner being.

As with any good teacher—and as with very few of the teachers I've known—my message has changed over time. I gave up on promoting civilization many years ago, even as I lived at the apex of empire in a large city in the interior southwestern United States. I later gave up on the idea that collapse of industrial civilization would allow me, my loved ones, or humans in general to persist long into the future. In light of this latter viewpoint, I have begun the transition to a Buddhism-inspired perspective. I explain this perspective with a brief story, probably apocryphal:

> *The Buddha asked one of his students how frequently he thought about death. The student replied, "Very often, probably dozens of time each day." The Buddha responded: "That's not enough. You must think about death with every breath."*

Life is urgent. The proverbial wolf is always at the door. I encourage people to conduct their lives with urgency while pointing out that pursuing and appreciating the present moment is what we have.

In short, I encourage people to live fully in the present. A few months ago I was asked for a three-sentence synopsis of an hour-long presentation I had just delivered. My response was the following: "How about three words? *Live here now.* If you really want three sentences, try these three: Live. Live now. Live here now."

I believe this is a decent way to pursue life even if all the models, data, assessments, and projections regarding climate change are incorrect. I encourage people to pursue what they love and also to pursue lives of excellence. I encouraged the latter pursuit throughout my career as a faculty member. Recently, though, I've added a dash of love and a sense of urgency to my message.

CHAPTER 2
ABRUPT CLIMATE CHANGE: GIVING NEW MEANING TO 'HARD' SCIENCE

BY GUY R. MCPHERSON

> *It is far better to grasp the universe as it really is than to persist in delusion, however satisfying and reassuring.*
>
> *~Carl Sagan~*

> *Facing the truth is so much easier than all the time and energy it takes running away from it.*
>
> *~Anonymous~*

American actress Lily Tomlin is credited with the expression, "No matter how cynical you become, it's never enough to keep up." With respect to climate science, our own efforts to stay abreast are blown away every week by new data, models, and assessments. It seems no matter how dire the situation becomes, it only gets worse when we check the latest reports.

Earth has never harbored humans at 3.5 C above baseline (i.e., the beginning of the Industrial Revolution, commonly accepted as 1750). I doubt a rapid rise to that temperature will leave habitat for humans on this planet. The ever-worsening bottom line with respect to greenhouse gases in the atmosphere indicates we've locked in 5 C temperature rise above baseline within the next few decades, according to an analysis conducted by U.K. climate scientist Peter Wasdell. Ignoring carbon dioxide and contemplating only methane bubbling from the Arctic Ocean takes us to 6 C above baseline by 2023 (and up to 16 C by 2033), according to Canadian climate scientist Paul Beckwith. Sam Carana's 1 April 2013 analysis of the same phenomenon leads to a global average temperature above 4 C by 2030 (and 10 C by 2040, http://methane-hydrates.blogspot.com/2013/04/methane-hydrates.html).

The response of politicians, heads of non-governmental organizations, and corporate leaders remains the same. They're paralyzed into inaction, and with good reason: there is no politically viable approach to dealing with climate change. As Halldor Thorgeirsson, a senior director with the United Nations Framework Convention on Climate Change, said on 17 September 2013, "We are failing as an international community. We are not on track (Thorgeirsson, 2013)." These are the people who know about, and presumably could do something about, our ongoing race to disaster (if only to sound the alarm). Tomlin's line is never more germane than when thinking about their pursuit of a buck at the expense of life on Earth.

Fully captured by corporations and the corporate states, the media continue to dance around the issue of climate change. Occasionally a forthright piece is published, but it generally points in the wrong direction, such as suggesting climate scientists and activists be killed (Delingpole, 2013). Leading mainstream outlets routinely misinform the public or report relevant information as if there is some question about the anthropogenic nature of climate change.

Worse than the media are the mainstream scientists who minimize the message at every turn. Scientists almost invariably underplay climate impacts. In some cases, scientists are aggressively muzzled by their governments. Science selects for conservatism. Academia selects for extreme conservatism. These folks are loath to risk drawing undue attention to themselves by pointing out that there might be a threat to civilization. Never mind the near-term threat to our entire species (in general, they couldn't care less about other species). If the truth is dire, they can find another not-so-dire version. The concept is supported by an article in the February 2013 issue of *Global Environmental Change* pointing out that climate change scientists routinely underestimate impacts "by erring on the side of least drama (Brysse, 2013)." Almost everybody reading these words has a vested interest in not wanting to think about climate change, which helps explain why the climate change deniers have won the battle. Like the rest of us, though, they've lost the war.

BEYOND LINEAR CHANGE

This common refrain is absurd: Earth can't possibly be responsive enough to climate change to make any difference to us. But, as the 27 May 2014 headline at the Skeptical Science website points out, "Rapid climate change is more deadly than asteroid impacts in Earth's past." That's correct. Abrupt climate change, which is fully under way, is more deadly than asteroids.

On 22 April 2014, *Truthout* correctly headlines their assessment, "Intergovernmental Climate Report Leaves Hopes Hanging on Fantasy Technology." Time follows up two days later with a desperate headline, "NASA Chief: Humanity's Future Depends On Mission To Mars" (NASA's lead administrator claims we need to occupy Mars by 2030).

"Attempts to reverse the impacts of global warming by injecting reflective particles into the stratosphere could make matters worse," according to research published in the 8 January 2014 issue of *Environmental Research Letters*. In addition, as described in the December 2013 issue of *Journal of Geophysical Research Atmospheres*, geo-engineering may succeed in cooling the Earth, but it would also disrupt precipitation patterns around the world. Furthermore, the "risk of abrupt and dangerous warming is inherent to the large-scale implementation of SRM," or solar radiation management, as pointed out in the 17 February 2014 issue of Environmental Research Letters. About a week later comes this line from research published in the 25 February 2014 issue of *Nature Communication*: "schemes to minimize the havoc caused by global warming by purposefully manipulating Earth's climate are likely to either be relatively useless or actually make things worse (Choi, 2014)." Finally, in a blow to technocrats, published online in the 25 June 2014 issue of *Nature Climate Change*, a large and distinguished group of international researchers concludes geo-engineering will not stop climate change (Barrett, 2014). As it turns out, the public isn't impressed, either. Research published in the 12 January 2014 issue of *Nature Climate Change* "reveals that the overall public evaluation of climate engineering is negative (Barrett, 2014)." Despite pervasive American ignorance about science, the public correctly interprets geo-engineering in the same light as the scientists, contrary to the techno-optimists.

The Intergovernmental Panel on Climate Change (IPCC) operates with a very conservative process and produces very conservative reports. Then governments of the world meddle with the reports to ensure Pollyanna outcomes, as reported by participants in the process (also see Nafeez Ahmed's 14 May 2014 report in *the Guardian*). David Wasdell's May 2014 analysis, which includes a critique of the IPCC's ongoing conservatism, concludes "equilibrium temperature increase predicted as a result of current concentration of atmospheric greenhouse gasses is already over 5°C (Wasdell, 2014)." I see no way for humans to survive such a rise in average global temperature.

Gradual change is not guaranteed, as pointed out by the U.S. National Academy of Sciences in December 2013: "The history of climate on the planet—as read in archives such as tree rings, ocean sediments, and ice cores—is punctuated with large changes that occurred rapidly, over the course of decades to as little as a few years (Hunziker, 2013)." The December 2013 report echoes one from Woods Hole Oceanographic Institution more than a decade earlier ("Abrupt Climate Change: Should We Be Worried?" 27 January 2003). Writing for the 3 September 2012 issue of *Global Policy*, Michael Jennings concludes, "a suite of amplifying feedback mechanisms, such as massive methane leaks from the sub-sea Arctic Ocean, have engaged and are probably unstoppable (Jennings, 2012)." During a follow-up interview with Alex Smith on Radio Ecoshock, Jennings admits "Earth's climate is already beyond the worst scenarios (Jennings, 2014)." *Truthout* states the following on 18 March 2014: "'climate change' is not the most critical issue facing society today; abrupt climate change is (Melton, 2014)." *Skeptical Science* finally catches up to reality on 2 April 2014 with an essay titled, "Alarming new study makes today's climate change more comparable to Earth's worst mass extinction (Lee, 2014)." The conclusion from this conservative source states the following: "Until recently the scale of the Permian Mass Extinction was seen as just too massive, its duration far too long, and dating too imprecise for a sensible comparison to be made with today's climate change. No longer (Lee, 2014)."

As reported by Robert Scribbler on 22 May 2014, "global sea surface temperature anomalies spiked to an amazing +1.25 degrees Celsius above the, already warmer than normal, 1979 to 2000 average. This departure is about 1.7 degrees C above 1880 levels—an extraordinary reading that signals the world may well be entering a rapid warming phase (Scribbler, 2014)."

Not to be outdone, now that abrupt climate change has entered the scientific lexicon, is dire news published in the 25

July 2014 issue of *Science*: "The study found that synchronization of the two regional systems began as climate was gradually warming. After synchronization, the researchers detected wild variability that amplified the changes and accelerated into an abrupt warming event of several degrees within a few decades (Praetorius, 2014)." Global average temperature rising "several degrees within a few decades" seems problematic to me, and to anybody else with a biological bent.

EXTINCTION OVERVIEW

On a planet 4°C hotter than baseline, all we can prepare for is human extinction, according to Oliver Tickell's 2008 synthesis in the *Guardian* (Tickell, 2008). Tickell is taking a conservative approach, considering humans have not been present at 3.5°C above baseline. According to the World Bank's 2012 report, "Turn down the heat: why a 4°C warmer world must be avoided (World Bank, 2012)," and an informed assessment of "BP Energy Outlook 2030" put together by Barry Saxifrage for the Vancouver Observer, our path leads directly to the 4°C mark. The Nineteenth Conference of the Parties of the UN Framework Convention on Climate Change (COP 19), held in November 2013 in Warsaw, Poland, was warned the following by Professor of Climatology Mark Maslin: "We are already planning for a 4°C world because that is where we are heading. I do not know of any scientists who do not believe that." Adding to planetary misery is a paper in the 16 December 2013 issue of the *Proceedings of the National Academy of*

Sciences concluding that 4°C above baseline terminates the ability of Earth's vegetation to sequester atmospheric carbon dioxide (University of Cambridge, 2012).

I'm not sure what it means to plan for 4°C (also known as extinction). I'm not impressed that civilized scientists claim to be planning for it, either. But I know we're human animals, and I know animals require habitat to survive. When there is no ability to grow food or secure potable water humans will exit the planetary stage. A rapid rise to 4°C takes us there.

According to Colin Goldblatt, author of a paper published online in the 28 July 2013 issue of *Nature Geoscience*, "The runaway greenhouse may be much easier to initiate than previously thought (Goldblatt, 2013)." Furthermore, as pointed out in the 1 August 2013 issue of *Science*, in the near term Earth's climate will change in orders of magnitude faster than at any time during the last sixty-five million years (Stanford, 2013). Tack on, without the large and growing number of self-reinforcing feedback loops reported since 2009, the 5°C rise in global average temperature fifty-five million years ago during a span of thirteen years, and it looks like trouble ahead for the wise ape. This conclusion ignores the long-lasting, incredibly powerful greenhouse gas discovered 9 December 2013 by University of Toronto researchers called Perfluorotributylamine (PFTBA). PFTBA is 7,100 times more powerful than carbon dioxide as a greenhouse gas in the atmosphere, and it persists for hundreds of years in the atmosphere (University of Toronto, 2013). It also ignores the irreversible nature of climate change:

Earth's atmosphere will harbor, at minimum, the current warming potential of atmospheric carbon dioxide concentration for at least the next thousand years, as indicated in the 28 January 2009 issue of the *Proceedings of the National Academy of Sciences*.

Finally, far too late, the *New Yorker* posits a relevant question on 5 November 2013: "Is It Too Late to Prepare for Climate Change (Kolbert, 2013)?" Joining the too-little, too-late gang, the Geological Society of London points out on 10 December 2013 that Earth's climate could be twice as sensitive to atmospheric carbon as previously believed (Summerhayes, 2013). *New Scientist*, in March 2014, points out that planetary warming is far more sensitive to atmospheric carbon dioxide concentration than indicated by past reports (LePage, 2014). As usual and expected, carbon dioxide emissions set a record again in 2013, the fifth-warming year on record and the second-warmest year without an El Nino.

IS THERE A WAY OUT?

All of the above information fails to include the excellent work of Tim Garrett, which points out that only complete collapse of industrial civilization avoids runaway greenhouse. Garrett reached the conclusion in a paper submitted in 2007 (personal communication) and published online by *Climatic Change* in November 2009 (Garrett, 2009). However, the outcry from civilized scientists delayed print publication until February 2011. The paper remains largely ignored by the scientific community, having been cited fewer than ten times since its publication.

According to Yvo de Boer, who was executive secretary of the United Nations Framework Convention on Climate Change in 2009, when attempts to reach a deal at a summit in Copenhagen crumbled with a rift between industrialized and developing nations, "the only way that a 2015 agreement can achieve a two-degree goal is to shut down the whole global economy (Morales, 2013) ." At least one politician finally caught up with Tim Garrett's excellent paper in Climatic Change.

Earth-system scientist Clive Hamilton concludes the following in his April 2013 book titled *Earthmasters*: "Without atmospheric sulfates associated with industrial activity Earth would be an extra 1.1 C warmer (Hamilton, 2013)." This estimate matches that of James Hansen and colleagues, who conclude 1.2 C cooling (plus or minus 0.2 C) as a result of atmospheric particulates (full paper in the 22 December 2011 issue of *Atmospheric Chemistry and Physics*). Both estimates are conservative relative to a paper in the 27 May 2013 issue of *Journal of Geophysical Research: Atmospheres*, which report a nearly 1 C temperature rise resulting from a 35-80% reduction in anthropogenic aerosols (Levy, 2013). In other words, collapse takes us directly to 2 C within a matter of weeks. Although 2 C has never been a scientific target, it has long been a political target. But even collapse of industrial civilization takes us to 2 C. Of course, maintaining industrial civilization takes us there, too.

Collapse of industrial civilization probably will save many non-human species, at least in the short term. Industrial civilization underlies the Sixth Great Extinction and the consequent loss of about 200 species each day. But the collapse of industrial civilization comes at a major cost to humans: it leads to the catastrophic meltdown of the world's nuclear power stations, which number more than 400. Considering the dire threat to humanity posed by Fukushima alone, it's difficult to imagine a situation in which a planet bathed in ionizing radiation would allow humans to persist many decades into the future.

Supporters of carbon farming—the nonsensical notion that industrial civilization can be used to overcome a predicament created by industrial civilization—claim all we need to do is fill the desert with non-native plants to the tune of an area three-quarters the size of the United States (Harball, 2013). And, they say, we'll be able to lower atmospheric carbon dioxide by a whopping 17.5 parts per million (ppm) in only two decades. Well, how exciting. At that blistering pace, atmospheric carbon dioxide will be all the way back down to the reasonably safe level of 280 ppm in only 140 years, more than a century after humans are likely to become extinct from climate change. And, based on research published in the 2 May 2014 issue of *Science*, soil carbon storage has been over estimated and is reduced as atmospheric carbon dioxide concentration rises (Groenigen, 2014).

According to the plan presented in the 23 August 2013 issue of *Scientific American*, the nonnative plants, irrigated with increasingly rare fresh water pumped by increasingly rare fossil-fuel energy, will sequester carbon sufficient to overcome contemporary emissions (Harball, 2013). Never mind the emissions resulting from pumping the water, or the desirability of converting thriving deserts into monocultures, or the notion of maintaining industrial civilization at the expense of non-civilized humans and non-human species. Instead, ponder one simple thought: when the nonnative plants die, they will emit back into the atmosphere essentially all the carbon they sequestered. A tiny bit of the carbon will be stored in the soil. The rest goes into the atmosphere as a result of decomposition.

TIPPED OVER

We crossed the planetary threshold in 2007 at about 0.76 C warming. At this point, according to David Spratt's excellent September 2013 report "Is Climate Already Dangerous? (Spratt, 2013)" not only had Arctic sea-ice passed its tipping point, but the Greenland ice sheet was not far behind, as the Arctic moves to sea-ice-free conditions in summer (the US Navy predicts an ice-free Arctic by summer 2016, a year later than expected by the United Kingdom Parliament, which points out that the six lowest September ice extents have occurred in the last six years, 2007-2012 (and now we can add 2013 to the list, and almost certainly 2014 as well).

Glaciologist Jason Box, an expert on Greenland ice, agrees. Box was quoted in a 5 December 2012 article in the *Guardian*: "In 2012, Greenland crossed a threshold where for the first time we saw complete surface melting at the highest elevations in what we used to call the dry snow zone. . . . As Greenland crosses the threshold and starts really melting in the upper elevations it really won't recover from that unless the climate cools significantly for an extended period of time which doesn't seem very likely (Goldenberg, 2012)." In January 2013, Box concluded we've locked in 69 feet—21 meters—of sea-level rise. Indeed, as stated in the September 2012 issue of *Global Policy*, "because of increasing temperatures due to greenhouse gas (GHG) emissions a suite of amplifying feedback mechanisms, such as massive methane leaks from the sub-sea Arctic Ocean, have engaged and are probably unstoppable (Jennings, 2012)." By December 2013, the disappearance of Greenland's ice had accelerated to five times the pace of a few years previously, and the IPCC was acknowledging they'd been far too conservative with past estimates. Continued conservatism is buttressed by research reported in the 16 March 2014 issue of *Nature Climate Change* (Khan, 2014) indicating the melting of Greenland ice accounts for about one-sixth of recent sea-level rise and also by research published in the 18 May 2014 issue of *Nature Geoscience* indicating (Morlighem, 2014) Greenland's icy reaches are far more vulnerable to warm ocean waters from climate change than had been thought.

Adding to the mounting evidence behind 2007 as the point of no return, Malcolm Light concluded on 22 December 2013, "We have passed the methane hydrate tipping point and are now accelerating into extinction as the methane hydrate 'Clathrate Gun' has begun firing volleys of methane into the Arctic atmosphere (Light, 2013)." According to Light's analysis in late 2013, the temperature of Earth's atmosphere will resemble that of Venus before 2100 (Light, 2013). Two weeks later, in an essay stressing near-term human extinction, Light concluded "The Gulf Stream transport rate started the methane hydrate (clathrate) gun firing in the Arctic in 2007 when its energy/year exceeded 10 million times the amount of energy/year necessary to dissociate subsea Arctic methane hydrates." Not surprisingly, the clathrate gun began firing in 2007, the same year the extent of Arctic sea ice reached a tipping point, according to David Spratt's analysis.

PREDICTING NEAR-TERM HUMAN EXTINCTION

If you think humans will adapt, think again. The rate of evolution trails the rate of climate change by a factor of 10,000, according to a paper in the August 2013 issue of Ecology Letters (Quintero, 2013). And it's not as if numerous extinction events haven't happened on this planet, as explained in the BBC program, *The Day the Earth Nearly Died*.

The rate of climate change clearly has gone beyond linear, as indicated by the presence of the myriad self-reinforcing feedback loops described below, and now threatens our species with extinction in the near term. Australian health officials added their voice to the discussion about extinction in an article titled, "Climate change could make humans extinct, warns health experts. The article was published 31 March 2014 in the Sydney Morning Herald, and the health officials clearly viewed 4 C as a problem to be dealt with later. Whatever we do now is too late." Anthropologist Louise Leakey ponders our near-term demise in her 5 July 2013 assessment at *Huffington Post* and her father Richard joins the fray in an interview a few months later (Leakey, 2013). Canadian wildlife biologist Neil Dawe joins the party of near-term extinction in an interview 29 August 2013 and musician-turned-activist Sir Bob Geldof joins the club in a *Daily Star* article from 6 October 2013. Australian health officials added their voices to the discussion about extinction in the 31 March 2014 issue of the *Sydney Morning Herald*, although they view 4 C as a problem to be dealt with later ("Climate change could make humans extinct, warns health expert"). Writing for *Truthout*, journalist John Feffer writes the following in his 27 April 2014 essay: "The planet and its hardier denizens may soldier on, but for us it will be game over (Feffer, 2014)." American linguist and philosopher Noam Chomsky concludes we're done in a 15 June 2014 interview with Chris Hedges at *Truthdig*, saying climate change "may doom us all, and not in the distant future (Hedges, 2014)." Larry Schwartz, writing for *AlterNet* on 21 July 2014, concludes, "Many environmentalists think we have already passed the point of no return (Schwartz, 2014)."

In the face of near-term human extinction, most Americans view the threat as distant and irrelevant, as illustrated by a 22 April 2013 article in the *Washington Post* based on poll results (Clement, 2013). These results echo the long-held sentiment that elected officials should be focused on the industrial economy, not far-away minor nuisances such as climate change. What if the citizenry knew how dire the situation has become? What if the mainstream media reported relevant information instead of constantly turning the issue over to future generations?

LARGE-SCALE ASSESSMENTS

During late 2007, Intergovernmental Panel on Climate Change's Fourth Assessment forecast a global average temperature rise in excess of 1.8 C by 2100. Depending on emissions scenarios, the rise in temperature is projected up to 4.5 C above baseline.

About a year later, the UK Hadley Centre for Meteorological Research projected about a 2 C rise in global average temperature by 2100. Later in 2008, the Hadley Centre's head of climate change predictions, Dr. Vicky Pope, called for a worst-case outcome of more than 5 C by 2100. Joe Romm, writing for *Grist*, claimed, "right now even Hadley understands it [> 5 C] is better described as the 'business-as-usual' case (Romm, 2008)."

In mid-2009, with data pouring in and modeling efforts becoming increasingly sophisticated, the United Nations Environment Programme projected a 3.5 C rise by 2100. By October 2009, the Hadley Centre for Meteorological Research forecast 4 C by 2060. The following month, Global Carbon Project and Copenhagen Diagnosis respectively projected 6 C and 7 C increases in temperature by 2100. Finally, the United

Nations Environment Programme projected up to 5 C by 2050 in its December 2010 analysis.

These assessments fail to account for significant self-reinforcing feedback loops (i.e., positive feedbacks, the term that implies the opposite of its meaning). The IPCC's vaunted Fifth Assessment continued the trend as it, too, ignored important feedbacks. As with prior reports, the Fifth Assessment "has been altered after the expert review stage, with changes added that downplay the economic impacts of a warming planet." Consider, for example, the failure to mention Arctic ice in the *Working Group Summary* released 31 March 2014.

On a positive note, major assessments fail to account for economic collapse. However, due to the four-decade lag between emissions and temperature rise, the inconvenient fact that the world has emitted more than twice the industrial carbon dioxide emissions since 1970 as we did from the start of the Industrial Revolution through 1970. Also, due to the self-reinforcing feedback loops described below, I strongly suspect it's too late for economic collapse to extend the run of our species. Bruce Melton, at *Truthout* in a 26 December 2013 piece featuring climate scientist Wallace Broeker, states the following:

> *Today we are operating on atmospheric concentrations of greenhouse gases from the 1970s. In the last 29 years we have emitted as many greenhouse gases as we emitted in the previous 236 years. Because of the great cooling effect of the oceans, we have not yet begun to see the warming*

that this recent doubling of greenhouse gases will bring (Melton, 2013).

Greenhouse gas emissions continue to accelerate even as the world's industrial economy slows to a halt. Emissions grew nearly twice as fast during the first decade of the new millennium as in the previous thirty years, as reported in the 11 April 2014 issue of the *Guardian* (Goldenberg, 2014).

The forty-year lag between cause and effect has been evident since at least 1938, when Guy Callendar pointed out the influence of rising carbon dioxide on global average temperature in a paper in the *Quarterly Journal of the Royal Meteorological Society*. A hand-drawn figure from Calendar's paper clearly shows a rise in global-average temperature beginning about 1915, roughly 40 years after the consumption of fossil fuels increased substantially. Callendar's work was used by J.S. Sawyer in a 1972 paper published in *Nature* to predict an "increase of 25% CO2 expected by the end of the century . . . [and] . . . an increase of 0.6°C in the world temperature" with stunning accuracy.

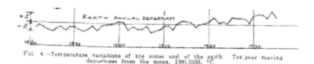

FIG. 4.—Temperature variations of the seas and of the earth. Ten year moving departures from the mean, 1901-1930, °C.

Astrophysicists have long believed Earth was near the center of the habitable zone for humans. Research published in the 10 March 2013 issue of *Astrophysical Journal* indicates Earth is on the inner edge of the habitable zone, and lies within 1% of inhabitability (one and a half million kilometers, or five times the distance from Earth to Earth's moon). A minor change in Earth's atmosphere removes human habitat. Unfortunately, we've invoked major changes in Earth's atmosphere, ramping up carbon dioxide by more than 40% and increasing methane a whopping 250% during the last 170 years.

The Northern Hemisphere is particularly susceptible to accelerated warming, as explained in the 8 April 2013 issue of *Journal of Climate*. Two days later a paper in Nature confirmed that summers in the Northern Hemisphere are hotter than they've been for six hundred years. As pointed out by Sherwood and Huber in the 25 May 2012 issue of the *Proceedings of the National Academy of Sciences* and then by James Hansen in his 15 April 2013 paper, humans cannot survive a wet-bulb temperature of 35 C (95 F).

As described by the United Nations Advisory Group on Greenhouse Gases in 1990, "Beyond 1 degree C may elicit rapid, unpredictable and non-linear responses that could lead to extensive ecosystem damage." James Hansen and crew finally caught up to the dire nature of 1 C warming, as reported in the 3 December 2013 issue of the *Guardian*. Unfortunately, their awakening came a full twenty-three years after the UN warning —twenty-eight self-reinforcing feedback loops too late.

We've clearly triggered the types of positive feedbacks the United Nations warned about in 1990. Yet my colleagues and acquaintances think we can and will work our way out of this horrific mess with the tools of industrial civilization (the ones that got us into this mess, as pointed out by Tim Garrett's excellent paper in the November 2009 issue of *Climatic Change*) or permaculture (which is not to denigrate permaculture, the principles of which are implemented at the Mud Hut). Reforestation doesn't come close to overcoming combustion of fossil fuels, as pointed out in the 30 May 2013 issue of *Nature Climate Change*. Furthermore, forested ecosystems do not sequester additional carbon dioxide as it increases in the atmosphere, as disappointingly explained in the 6 August 2013 issue of *New Phytologist*.

Some green-washing solutionistas take refuge in the nuclear solution. It's astonishing what one can conclude when grid-tied electricity is viewed as a natural right. James Hansen's endorsement notwithstanding, nuclear power plants cause rather than prevent additional warming of Earth, in large part because the carbon emissions associated with concrete are extremely high. In addition, Three Mile Island, Chernobyl, and Fukushima forever buried the notion of clean nuclear energy.

Much of the story behind abrupt climate change can be told by referring to two long-forgotten documents. First, a paper commissioned by the Pentagon and written by consultants Peter

Schwartz and Doug Randall in October 2003 ("An Abrupt Climate Change Scenario and Its Implications for United States National Security") includes this line: "[A]n abrupt climate change scenario could potentially de-stabilize the geo-political environment, leading to skirmishes, battles, and even war due to resource constraints."

Six years after the Pentagon report came a briefing from the Association of Small Island States prepared for the 2009 United Nations Conference of the Parties in Copenhagen (COP15). The briefing's summary contains this statement: "THE LONG-TERM SEA LEVEL THAT CORRESPONDS TO CURRENT CO2 CONCENTRATION IS ABOUT 23 METERS ABOVE TODAY'S LEVELS, AND THE TEMPERATURES WILL BE 6 DEGREES C OR MORE HIGHER. THESE ESTIMATES ARE BASED ON REAL LONG TERM CLIMATE RECORDS, NOT ON MODELS."

In other words, extinction of humans was already guaranteed, to the knowledge of Obama and his administration. Even before the scientific community reported the dire feedback, the administration abandoned climate change as a significant issue because it knew we were headed for extinction as early as 2009. Rather than shoulder the unenviable task of truth-teller, Obama lied about the gravity of climate change. And he still does.

Ah, those were the good ol' days, back when atmospheric carbon dioxide concentrations were well below 400 ppm. We've recently blown through the 400 ppm mark for the first time in the human experience. As reported in the journal *Global and Planetary Change* in April 2013, every molecule of atmospheric carbon dioxide since 1980 comes from human emissions. Tacking on a few of the additional greenhouse gases contributing to climate change and taking a conservative approach jacks up the carbon dioxide equivalent to more than 480 ppm.

SELF-REINFORCING FEEDBACK LOOPS

Self-reinforcing feedback loops, or positive feedbacks, "feed" upon themselves. Whereas a soccer ball kicked across a soccer field quickly slows to a halt, the same ball kicked over a cliff actually picks up speed as it falls. Industrial civilization has triggered many such feedbacks. They are described below in the order presented by the scientific community. But first, a few introductory words about methane are warranted.

Seeps of methane are appearing in numerous locations off the eastern coast of the United States, leading to rapid destabilization of methane hydrates (according to the 25 October 2013 issue of Nature). On land, anthropogenic emissions of methane in the United States have been severely underestimated by the Environmental Protection Agency, according to a paper in the 25 November 2013 issue of *Proceedings of the National Academy of Sciences*. This figure is 1,100 parts per billion (ppb) higher than pre-industrial peak levels. Methane release tracks closely with temperature rise throughout Earth history— specifically, Arctic methane release and rapid global temperature rise are interlinked—including a temperature rise up to about 1 C per year over a decade, according to data from ice cores. The tight linkage between Arctic warming and planetary warming was verified in an article in the 2 February 2014 issue of *Nature Geoscience*, which found that the Arctic's cap of cold, layered air plays a more important role in boosting polar warming than does its shrinking ice and snow cover. A layer of shallow, stagnant air acts like a lid, concentrating heat near the surface. Finally, adding fuel to the growing fire, a paper in the 27 March 2014 issue of *Nature* articulates the strong interconnection between methane release and temperature rise: "For each degree that Earth's temperature rises, the amount of methane entering the atmosphere . . . will increase several times. As temperatures rise, the relative increase of methane emissions will outpace that of carbon dioxide."

Self-reinforcing feedback loops are listed and briefly described below. As with the other information in this chapter, updates can be found at Nature Bats Last (McPherson, 2009).

1. Methane hydrates are bubbling out of the Arctic Ocean (*Science*, March 2010). As described in a subsequent paper in the June 2010 issue of *Geophysical Research Letters*, a minor increase in temperature would cause the release of upwards of 16,000 metric tons of methane each year. Storms accelerate the release, according to research published in the 24 November 2013 issue of *Nature Geoscience*. According to NASA's CARVE project, methane plumes were up to

150 kilometers across as of mid-July 2013. Global average temperature is expected to rise by more than 4 C by 2030 and 10 C by 2040 based solely on methane release from the Arctic Ocean, according to Sam Carana's research at *Arctic News*. Whereas Malcolm Light's 9 February 2012 forecast of extinction of all life on Earth by the middle of this century, published by *Arctic News*, appeared premature because his conclusion of exponential methane release during summer 2011 was based on data subsequently revised and smoothed by US government agencies, subsequent information—most notably from papers published in *Science, Geophysical Research Letters, Nature Geoscience*, and also by NASA's CARVE project—indicates the grave potential for catastrophic release of methane. Industrial civilization appears incapable of killing all life on Earth, although that clearly is one of its goals.

Catastrophically rapid release of methane in the Arctic is further supported by Nafeez Ahmed's thorough analysis in the 5 August 2013 issue of the *Guardian* as well as Natalia Shakhova's 2012 interview with Nick Breeze at the European Geophysical Union in Vienna. In early November 2013, methane levels well in excess of 2,600 ppb were recorded at multiple altitudes in the Arctic. Later that same month, Shakhova and colleagues published a paper in *Nature Geoscience* suggesting "significant quantities of methane are escaping the East Siberian Shelf" and indicating that a fifty-billion-tonne "burst" of methane could warm Earth by 1.3 C. Such a burst of methane is "highly possible at any time." By 15 December 2013, methane bubbling up from the seafloor of the Arctic Ocean had sufficient force to prevent sea ice from forming in the area.

The Venus syndrome elucidated by Sam Carana and Malcolm Light a few pages back finally is validated by the refereed journal literature. An article in the 3 February 2014 issue of *Journal of Geophysical Research: Earth Surface* claiming, "Sustained submergence into the future should increase gas venting rate roughly exponentially as sediments continue to warm."

The importance of methane cannot be overstated. Increasingly, evidence points to a methane burst underlying the Great Dying associated with the end-Permian extinction event, as pointed out in the 31 March 2014 issue of *Proceedings of the National Academy of Sciences*. As Malcolm Light reported on 14 July 2014: "There are such massive reserves of methane in the subsea Arctic methane hydrates, that if only a few percent of them are released, they will lead to a jump in the average temperature of the Earth's atmosphere of 10 degrees C and produce a 'Permian' style major extinction event which will kill us all."

Discussion about methane release from the Arctic Ocean has been quite heated (pun intended). Paul Beckwith was criticized by the conservative website

Skeptical Science. His response at *Arctic News* from 9 August 2013 includes the following statement: "[T]he climate system has entered a period of abrupt change that has not been seen before in human history, but has happened many times in the paleorecords. In fact, rates of change now are at least 10x higher than any seen in the geologic record."

Robert Scribbler provides a terrifying summary dated 24 February 2014, and concludes, "two particularly large and troubling ocean to atmosphere methane outbursts were observed" in the Arctic Ocean. Such an event hasn't occurred during the last forty-five million years. The following is Scribbler's bottom line: "That time of dangerous and explosive reawakening, increasingly, seems to be now."

2. Warm Atlantic water is defrosting the Arctic as it shoots through the Fram Strait, as reported in the January 2011 issue of *Science*. Extent of Arctic sea ice passed a tipping point in 2007, according to research published in the February 2013 issue of *The Cryosphere*. Subsequent melting of Arctic ice is reducing albedo, hence enhancing absorption of solar energy. "Averaged globally, this albedo change is equivalent to 25% of the direct forcing from CO2 during the past 30 years," according to research published in the 17 February 2014 issue of the *Proceedings of the National Academy of Sciences.*

Destabilization of the deep circulation in the Atlantic Ocean may be "spasmodic and abrupt rather than a more gradual increase" as earlier expected, according to a paper published in the 21 February 2014 issues of *Science*. Models continue to underestimate relative to observations, as reported in the 10 March 2014 issue of *Geophysical Research Letters*.

3. Siberian methane vents have increased in size from less than a meter across in the summer of 2010 to about a kilometer across in 2011, as reported in *Tellus*'s February 2011 issue. According to a paper in the 12 April 2013 issue of *Science*, a major methane release is almost inevitable, which makes us wonder where the authors have been hiding. Almost inevitable, they report, regarding an ongoing event. *National Geographic* reported in its 17 April 2014 issue that trees are tipping over and dying as permafrost thaws, thus illustrating how self-reinforcing feedback loops feed each other. Meanwhile, several large holes were discovered in Siberia during the summer of 2014. The initial reaction from a refereed journal, in this case a paper published in the 31 July 2014 issue of Nature, indicates atmospheric methane levels more than 50,000 times the usual level.

4. Peat in the world's boreal forests is decomposing at an astonishing rate, according to research reported in the November 2011 of *Nature Communications*.

5. The March 2012 issue of *Environmental Research Letters* includes a description of tall shrubs invading new areas. The attendant deep roots destabilize the permafrost.

6. Greenland ice is darkening, as reported in the June 2012 issue of *The Cryosphere*.

7. Methane is being released from the Antarctic, too, according to research published in the August 2012 issue of *Nature*. According to a paper in the 24 July 2013 issue of *Scientific Reports*, melt rate in the Antarctic has caught up to the Arctic. The West Antarctic Ice Sheet (WAIS) is losing over 150 cubic kilometers of ice each year according to CryoSat observations published 11 December 2013, and Antarctica's crumbling Larsen B Ice Shelf is poised to finish its collapse, according to Ted Scambos, a glaciologist at the National Snow and Ice Data Center at the annual meeting of the American Geophysical Union. Loss of Antarctic ice is accelerating even in areas long considered stable, as documented in the 24 July 2014 edition of *Scientific Reports*. The rate of loss from 2010 to 2013 was double that during the years 2005 to

2010, according to the online version of a paper posted 16 June 2013 in *Geophysical Research Letters*.

8. NASA reported on expanding Russian forest and bog fires in August 2012. The phenomenon was consequently apparent throughout the Northern Hemisphere according to the July 2013 issue of *Nature Communications*. The *New York Times* reported hotter, drier conditions leading to huge fires in the western part of North America as the "new normal" in their 1 July 2013 issue. A paper in the 22 July 2013 issue of the *Proceedings of the National Academy of Sciences* indicated boreal forests are burning at a rate exceeding that of the last 10,000 years. According to reports from Canada's Interagency Fire Center, total acres burned to date in early summer 2014 are more than six times that of a typical year. This rate of burning is unprecedented not just for this century, but for any period in Canada's basement forest record over the last 10,000 years.

9. Cracking of glaciers accelerates in the presence of increased carbon dioxide, according to a paper in the October 2012 issue of *Journal of Physics D: Applied Physics*.

10. The Beaufort Gyre apparently has reversed course, based on an October 2012 report by the U.S. National

Snow and Ice Data Center. No subsequent evidence confirms this reversal. Mechanics of this process are explained by the Woods Hole Oceanographic Institution in a report published online 14 February 2014

11. Exposure to sunlight increases bacterial conversion of exposed soil carbon, thus accelerating thawing of the permafrost, as reported in the February 2013 issue of *Proceedings of the National Academy of Sciences*. Subsequent carbon release "could be expected to more than double overall net C losses from tundra to the atmosphere," as reported in the March 2014 issue of *Ecology*. Arctic permafrost houses about half the carbon stored in Earth's soils, an estimated 1,400 to 1,850 petagrams of it, according to NASA. Peat chemistry changes as warming proceeds, which accelerates the process, as reported in the 7 April 2014 issue of *Proceedings of the National Academy of Sciences*.

12. The microbes have joined the party, too, according to a paper in the 23 February 2013 issue of *New Scientist*

13. As reported in the April 2013 issue of *Nature Geoscience*, summer ice melt in Antarctica is at its highest level in a thousand years; summer ice in the Antarctic is melting ten times quicker than it was 600 years ago, with the most rapid melt occurring in the last fifty years.

According to a paper in the 4 March 2014 issue of *Geophysical Research Letters*—which assumes relatively little change in regional temperature during the coming decades—"modeled summer sea-ice concentrations decreased by 56% by 2050 and 78% by 2100." Citing forthcoming papers in *Science* and *Geophysical Research Letters*, the 12 May 2014 issue of the *New York Times* reported the following:

> *A large section of the mighty West Antarctica Ice Sheet has begun falling apart and its continued melting now appears to be unstoppable. . . . The new finding appears to be the fulfillment of a prediction made in 1978 by an eminent glaciologist, John H. Mercer of the Ohio State University. He outlined the vulnerable nature of the West Antarctic Ice Sheet and warned that the rapid human-driven release of greenhouse gases posed 'a threat of disaster.'*

Although scientists have long expressed concern about the instability of the West Antarctic Ice Sheet (WAIS), a research paper published in the 28 August 2013 of *Nature* indicates the East Antarctic Ice Sheet (EAIS) has undergone rapid changes in the past five decades. The latter is the world's largest ice sheet and was previously thought to be at little risk from climate change. Now, the melting of the EAIS is signaling a potential threat to global sea levels. The EAIS holds enough water to raise sea levels more than fifty meters.

14 Increased temperature and aridity in the southwestern interior of North America facilitates movement of dust from low-elevation deserts to high-elevation snowpack, thus accelerating snowmelt, as reported in the 17 May 2013 issue of *Hydrology and Earth System Sciences*.

15. Floods in Canada are sending pulses of silty water out through the Mackenzie Delta and into the Beaufort Sea, thus painting a wide section of the Arctic Ocean near the Mackenzie Delta brown (from NASA, June 2013). Pictures of this phenomenon are shown on NASA's website.

16. Surface meltwater draining through cracks in an ice sheet can warm the sheet from the inside, softening the ice and letting it flow faster, according to a study published in the September 2013 issue of *Geophysical Research: Earth Surface*. It appears a Heinrich Event has been triggered in Greenland. Consider the description of such an event as provided by Robert Scribbler on 8 August 2013:

> *In a Heinrich Event, the melt forces eventually reach a tipping point. The warmer water has greatly*

softened the ice sheet. Floods of water flow out beneath the ice. Ice ponds grow into great lakes that may spill out both over top of the ice and underneath it. Large ice damns (sic) may or may not start to form. All through this time ice motion and melt is accelerating. Finally, a major tipping point is reached and in a single large event or ongoing series of such events, a massive surge of water and ice flush outward as the ice sheet enters an entirely chaotic state. Tsunamis of melt water rush out bearing their vast floatillas (sic) of ice burgs (sic), greatly contributing to sea level rise. And that's when the weather really starts to get nasty. In the case of Greenland, the firing line for such events is the entire North Atlantic and, ultimately the Northern Hemisphere.

17. Breakdown of the thermohaline conveyor belt is happening in the Antarctic as well as the Arctic, thus leading to melting of Antarctic permafrost, according to research published in the July 2013 issue of *Scientific Reports*. In the past sixty years, the ocean surface offshore Antarctica became less salty as a result of melting glaciers and more precipitation, as reported in the 2 March 2014 issue of *Nature Climate Change*.

18. Loss of Arctic sea ice is reducing the temperature gradient between the poles and the equator, thus causing the jet stream to slow and meander (ongoing research by

Rutgers University's Jennifer Francis is particularly relevant). The most extreme "dipole" on record occurred during 2013-2014, as reported in 16 May 2014 issue of *Geophysical Research Letters*. One result is the creation of weather blocks such as the very high temperatures in Alaska during 2013 and 2014. This so-called "polar vortex" became widely reported in the United States in 2013 and received the attention of the academic community when the 2013-2014 drought threatened crop production in California.

One result of this phenomenon is boreal peat drying and catching fire like a coal seam. The resulting soot enters the atmosphere to fall again, coating the ice surface elsewhere, thus reducing albedo and hastening the melting of ice. Each of these individual phenomena has been reported, albeit rarely. The inability or unwillingness of the media to connect two dots is not surprising, and has been routinely reported. Extreme weather events are occurring, as reported in the 22 June 2014 issue of *Nature Climate Change*.

19. Arctic ice is growing darker, hence less reflective, as reported in the August 2013 issue of *Nature Climate Change.*

20. Extreme weather events drive climate change, as reported in the 15 August 2013 issue of *Nature.*

21. Drought-induced mortality of trees contributes to increased decomposition of carbon dioxide into the atmosphere and decreased sequestration of atmospheric carbon dioxide. Such mortality has been documented throughout the world since at least November 2000 in *Nature*, with recent summaries in the February 2013 issue of Nature for the tropics and in the August 2013 issue of *Frontiers in Plant Science* for temperate North America.

One extremely important example of this phenomenon is occurring in the Amazon, where drought in 2010 led to the release of an enormous amount of carbon (as reported in *Science*, February 2011). The conventional calculation badly underestimates the carbon release, according to research published in the 27 May 2014 online issue of *Global Change Biology*. In addition, ongoing deforestation in the region is driving declines in precipitation at a rate much faster than long thought, as reported in the 19 July 2013 issue of *Geophysical Research Letters*. An overview of the phenomenon, focused on the Amazon, was provided by *Climate News Network* on 5 March 2014.

Tropical rainforests, long believed to represent the primary driver of atmospheric carbon dioxide, are on the verge of giving up that role. According to a 21 May 2014 paper published in Nature, "the higher turnover rates of carbon pools in semi-arid biomes are an increasingly important driver of global carbon cycle inter-annual variability," indicating the emerging role of drylands in controlling environmental conditions.

22. Ocean acidification leads to release of less dimethyl sulfide (DMS) by plankton. DMS shields Earth from radiation, as reported in the 25 August 2013 issue of *Nature Climate Change*. Plankton form the base of the marine food web, and are on the verge of disappearing completely, according to a paper in the 17 October 2013 issue of *Global Change Biology*. As with carbon dioxide, ocean acidification is occurring rapidly, according to a paper in the 26 March 2014 issue of *Global Biogeochemical Cycles*. Acidification is proceeding at a pace unparalleled during the last 300 million years, according to research published in the 2 March 2012 issue of *Science*.

23. Jellyfish have assumed a primary role in the oceans of the world (26 September 2013 issue of the *New York Times Review of Books*, in a review of Lisa-Ann Gershwin's book, *Stung! On Jellyfish Blooms and the Future of the Ocean*). Gershwin states, "We are creating a world more like the late Precambrian than the late 1800s— a world where jellyfish ruled the seas and organisms with shells didn't exist. We are creating a world where we humans may soon be unable to survive, or want to."

Jellyfish contribute to climate change via (1) release of carbon-rich feces and mucus used by bacteria for respiration, thereby converting bacteria into carbon dioxide factories and (2) consumption of vast numbers of copepods and other plankton.

24. Sea-level rise causes slope collapse, tsunamis, and release of methane, as reported in the September 2013 issue of *Geology*. In eastern Siberia, the speed of coastal erosion has nearly doubled during the last four decades as the permafrost melts.

25. As reported in the September 2013 issue of *Nature Climate Change,* rising ocean temperatures will upset natural cycles of carbon dioxide, nitrogen and phosphorus, hence reducing plankton.

26. Earthquakes trigger methane release, and consequent warming of the planet triggers earthquakes, as reported by Sam Carana at *Arctic News* during October 2013.

27. Small ponds in the Canadian Arctic are releasing far more methane than expected based on their aerial cover, as reported in *PLoS ONE's* November 2013 issue. This is the first of several freshwater ecosystems releasing methane into the atmosphere, as reviewed in the 19 March 2014

issue of *Nature* and subsequently described by a large-scale study in the 28 April 2014 issue of *Global Change Biology*.

28. Mixing of the jet stream is a catalyst, too. High methane releases follow fracturing of the jet stream, accounting for past global average temperature rises up to 16 C in a decade or two. This phenomenon was reported by University of Ottawa climate scientist Paul Beckwith via video on 19 December 2013.

28. As research indicates, "fewer clouds form as the planet warms, meaning less sunlight is reflected back into space, driving temperatures up further still." This phenomenon solves a long-puzzling question among climate scientists, and was reported in the 2 January 2014 issue of *Nature*.

30. "Thawing permafrost promotes microbial degradation of cryo-sequestered and new carbon leading to the biogenic production of methane," as reported in *Nature Communications* on 14 February 2014.

31. As reported in the 27 February 2014 issue of *Nature Geoscience*, a natural, invisible hole extends over several thousand kilometers in a layer that prevents transport of most of the natural and human-made substances into the

stratosphere by virtue of its chemical composition. This hole extends over the tropical West Pacific. Like in a giant elevator, many chemical compounds emitted at the ground pass thus unfiltered through this so-called "detergent layer" of the atmosphere. Global methane emissions from wetlands are currently about 165 terragrams (megatons metric) each year. This research estimates that annual emissions from these sources will increase by up to 260 megatons annually. By comparison, the total annual methane emission from all sources (including the human addition) is about 600 megatons each year.

32. Deep ocean currents are apparently slowing. According to one of the authors of the paper published in the 2 March 2014 issue of *Nature Climate Change*, "we're likely going to see less uptake of human produced, or anthropogenic, heat and carbon dioxide by the ocean, making this a positive feedback loop for climate change." Because this phenomenon contributed to cooling and sinking of the Weddell polynya, "it's always possible that the giant polynya will manage to reappear in the next century. If it does, it will release decades' worth of heat and carbon from the deep ocean to the atmosphere in a pulse of warming." Subsequent model results indicate "large spatial redistribution of ocean carbon," as reported in the March 2014 issue of the *Journal of Climate*.

33. As reported in the 2 May 2014 issue of *Science*, increased atmospheric carbon dioxide causes soil microbes to produce more carbon dioxide.

34. Reductions in seasonal ice cover in the Arctic "result in larger waves, which in turn provide a mechanism to break up sea ice and accelerate ice retreat" (*Geophysical Research Letters, 5 May 2014*)

35. A huge hidden network of frozen methane and methane gas, along with dozens of spectacular flares firing up from the seabed, has been detected off the North Island of New Zealand. These preliminary results were reported in the 12 May 2014 issue of the *New Zealand Herald*. The first evidence of widespread active methane seepage in the Southern Ocean, off the sub-Antarctic island of South Georgia, was subsequently reported in the 1 October 2014 issue of *Earth and Planetary Science Letters*.

36. As reported in the 8 June 2014 issue of *Nature Geoscience*, rising global temperatures could increase the amount of carbon dioxide naturally released by the world's oceans, fuelling further climate change.

37. As global average temperature increases, "the concentrations of water vapor will also increase in

response to that warming. This moistening of the atmosphere, in turn, absorbs more heat and further raises the Earth's temperature." (*Proceedings of the National Academy of Science*, 28 July 2014)

38. Arctic drilling was fast-tracked by the Obama administration during the summer of 2012.

39. Supertankers are taking advantage of the slushy Arctic, demonstrating that every catastrophe represents a business opportunity, as pointed out by Professor of Journalism Michael I. Niman and picked up by *Truthout* during September 2013.

As nearly as I can distinguish, only the latter two feedback processes are reversible at a temporal scale relevant to our species. Once you pull the tab on the can of beer, there's no keeping the carbon dioxide from bubbling up and out. These feedbacks are not additive, they are multiplicative; they not only reinforce within a feedback, the feedbacks also reinforce among themselves. Now that we've entered the era of expensive oil, I can't imagine we'll voluntarily terminate the process of drilling for oil and gas in the Arctic (or anywhere else). We also will not willingly forgo a few dollars by failing to take advantage of the long-sought Northwest Passage or make any attempt to slow economic growth.

Never mind that American naturalist George Perkins Marsh predicted anthropogenic climate change as a result of burning fossil fuels in 1847. Never mind the warning issued by filmmaker Frank Capra in 1958 or the one issued by Austrian philosopher Ivan Illich in his 1973 article in *Le Monde*: "The impact of industrially packaged quanta of energy on the social environment tends to be degrading, exhausting, and enslaving, and these effects come into play even before those which threaten the pollution of the physical environment and the extinction of the (human) race." Never mind the warning and plug for geo-engineering issued by US President Lyndon B. Johnson's Science Advisory Committee in 1965 that stated "the climate changes that may be produced by the increased CO_2 content could be deleterious from the point of view of human beings. The possibilities of deliberately bringing about countervailing climatic changes therefore need to be thoroughly explored." Never mind the 1986 warning from NASA's Robert Watson of "human misery in a few decades" and eventual human extinction as a result of climate change. Never mind that climate risks have been underestimated for the last 20 years, or that the IPCC's efforts have failed miserably (as indicated by David Wasdell's scathing indictment of the vaunted Fifth Assessment for the Apollo-Gaia Project in February 2014). After all, climate scientist Kevin Anderson tells us what I've known for years: politicians and the scientists writing official reports on climate change are lying, and we have less time than most people can imagine. (Consider the minor example of the US Environmental Protection Agency "underestimating" by one hundred to one thousand times the methane release associated with hydro-fracturing to extract natural gas, as reported in the 14 April 2014 issue of the *Proceedings of the National Academy of Sciences*.) Never mind David Wasdell pointing out in 2008 that we must have a period of negative radiative forcing merely to end up with a stable, non-catastrophic climate system. Never mind that Peruvian ice requiring 1,600 years to accumulate has melted in the last twenty-five years, according to a paper in the 4 April 2013 issue of *Science*. And never mind that summer warming in the interior of large continents in the Northern Hemisphere has outstripped model predictions in racing to 6-7 C since the last Glacial Maximum, according to a paper that tallies temperature rise in China's interior in the 15 May 2013 issue of the *Proceedings of the National Academy of Sciences*. And finally, never mind that the IPCC's projections have been revealed as too conservative time after time, including low-balling the impact of emissions, as pointed out in the 9 March 2014 issue of *Nature Climate Change*. On 24 March 2014, renowned climate scientist Michael Mann commented on climate change as reported in the IPCC's Fifth Assessment, stating "It's not far-off in the future and it's not exotic creatures—it's us and now." As the Fifth Assessment admits, climate change has already left its mark "on all continents and across the oceans."

Never mind all that. Future temperatures will likely be at the higher end of the projected range because the forecasts are all too conservative and climate negotiations won't avert catastrophe. They won't avert catastrophe because there is no politically viable approach to dealing with climate change. I cannot imagine a politician running on the campaign of complete collapse of industrial civilization, which we've known as the only "solution" to prevent runaway climate change since Tim Garrett's classic paper was published in 2009.

Catastrophe is under way. Global oceans have risen approximately ten millimeters per year during the last two years. This rate of rise is more than three times the rate of sea-level rise during the time of satellite-based observations from 1993 to the present. Ocean temperatures are rising, and have been impacting global fisheries for four decades, according to the 16 May 2013 issue of *Nature*.

Catastrophe is not widely distributed in the United States. Well, not yet, even though the continental US experienced its highest temperature ever in 2012, shattering the 1998 record by a full degree Fahrenheit. The eastern coast of North America experienced its hottest water temperatures all the way to the bottom of the ocean during the same period. The epic dust bowl of 2012 grew and grew and grew all summer long. As pointed out in the March 2004 issue of *Geophysical Research Letters*, disappearing sea ice is expectedly contributing to the drying of the western United States (more definitive research on the topic appeared in the December 2005 issue of *Earth Interactions*). Equally expectedly, the drought arrived forty years earlier than anticipated.

Even James Hansen and Makiko Sato are asking whether the loss of ice on Greenland has gone exponential (while ridiculously calling for a carbon tax to "fix" the "problem"), and the tentative answer is not promising, based on very recent data, including a nearly five-fold increase in melting of Greenland's ice since the 1990s and a stunning melting of ninety-eight percent of Greenland's ice surface between 8 and 15 July 2012. The mainstream media are taking notice, with the 18 July 2013 issue of *Washington Post* reporting the ninth highest April snow cover in the Northern Hemisphere giving way to the third lowest snow cover on record the following month (relevant records date to 1967, and the article is headlined, "Snow and Arctic sea ice extent plummet suddenly as globe bakes").

On a particularly dire note for humans, climate change causes the early death of five million people each year. Adding to our misery are interactions between various aspects of environmental decay. For example, warming in the Arctic is causing the release of toxic chemicals long trapped in the region's snow, ice, ocean, and soil, according to research published in the 24 July 2011 issue of *Nature Climate Change.*

Greenhouse gas emissions keep rising, and keep setting records. According to a 10 June 2013 report by the International Energy Agency, the horrific trend continued in 2012 when carbon dioxide emissions set a record for the fifth consecutive year. The trend puts disaster in the cross hairs, with the ever-conservative International Energy Agency claiming we're headed for a temperature in excess of 5 C.

In 2013, the US State of the Climate published the following statements on 17 July 2014 as a supplement to the July 2014 issue of the *Bulletin of the American Meteorological Society*:

- Ocean surface continues to warm

- Sea levels reach a record high

- Glaciers retreat for the twenty-fourth consecutive year

- Greenhouse gases continue to climb

- The planet's surface remains near its warmest

- Warm days are increasing and cool nights are decreasing

Completely contrary to the popular contrarian myth, global warming has accelerated, with more overall global warming in the fifteen years up to March 2013 than the prior fifteen years. This warming has resulted in about ninety percent of overall global warming going into heating the oceans, and the oceans have been warming dramatically, according to a paper published in the March 2013 issue of *Geophysical Research Letters*. A paper in the 20 March 2014 issue of *Environmental Research Letters* points out that land-surface temperatures poorly measure global warming. About thirty percent of the ocean warming over the past decade has occurred in the deeper oceans below 700 meters, which is unprecedented over at least the past half century. According to a paper in the 1 November 2013 issue of *Science*, the rate of warming of the Pacific Ocean during the last sixty years is fifteen times faster than at any time during the last 10,000 years. By the end of 2013, the fourth hottest year on record, the deep oceans were warming particularly rapidly and NASA and NOAA reported no pause in the long-term warming trend. "In 2013 ocean warming rapidly escalated, rising to a rate in excess of 12 Hiroshima bombs per second—over three times the recent trend." When the heat going into the ocean begins to influence land-surface temperatures, "rapid warming is expected," according to a paper published 9 February 2014 in *Nature Climate Change*. According to James Wight, writing for *Skeptical Science* on 12 March 2014, "Earth is gaining heat faster than ever."

Coincident with profound ocean warming, the death spiral of Arctic sea ice is well under way. As reported in the 22 February 2014 issue of *Geophysical Research Letters*, sea-surface temperatures have increased 0.5 to 1.5 C during the last decade. "The seven lowest September sea ice extents in the satellite record have all occurred in the past seven years."

In the category of myth busting comes recent research published in the August 2013 issue of *Proceedings of the National Academy of Sciences*. Contrary to the notion that changing solar radiation is responsible for rising global temperature, the amount of solar radiation passing through Earth's atmosphere and reaching the ground globally peaked in the 1930s, substantially decreased from the 1940s to the 1970s, and changed little after that. Indeed, the current solar activity cycle is the weakest in a century. In addition, according to a paper in the 22 December 2013 issue of *Nature Geoscience*, climate change has not been strongly influenced by variations in heat from the sun.

Global loss of sea ice matches the trend in the Arctic. It's down, down, and down some more, with the five lowest values on record all happening in the last seven years (through 2012). As reported in a June 2013 issue of *Science*, the Antarctic's ice shelves are melting from below. When interviewed for the associated article in the 13 June 2013 issue of *National Geographic*, scientists expressed surprise at the rate of change. Three months later the 13 September 2013 issue of *Science* contains another surprise for mainstream scientists. The Pine Island Glacier is melting from below as a result of warming seawater. And four months after that dire assessment the massive glacier was melting irreversibly, according to a paper in the 12 January 2014 issue of *Nature Climate Change*.

THEN SEE WHERE WE'RE GOING

The climate situation is much worse than we've been led to believe, and is accelerating far more rapidly than accounted for by models. Even the US Centers for Disease Control and Prevention acknowledges, in a press release dated 6 June 2013, potentially lethal heat waves on the near horizon. A month later the World Meteorological Organization pointed out that Earth experienced unprecedented recorded climate extremes during the 2001-2010 decade, contributing to more than a 2,000% increase in heat-related deaths.

Although climate change's heat—not cold—is the real killer, according to research published in the December 2013 issue of the Journal of Economic Literature, swings in temperature may be even more lethal than high temperatures. Specifically, research published in the 29 January 2014 issue of the *Proceedings of the Royal Society of London* indicates insects are particularly vulnerable to temperature swings.

Ice sheet loss continues to increase at both poles, and warming of the West Antarctic Ice Sheet is twice the earlier scientific estimate. Arctic ice is at an all-time low, half that of 1980, and the Arctic has lost enough sea ice to cover Canada and Alaska in 2012 alone. In short, summer ice in the Arctic is nearly gone. Furthermore, the Arctic could well be free of ice by September 2015 or 2016, an event that last occurred some three million years ago, long before the genus Homo walked the planet. Among the consequences of declining Arctic ice is extremes in cold weather in northern continents (thus illustrating why "climate change" is a better term than "global warming").

The Eemian interglacial period that began some 125,000 years ago is often used as a model for contemporary climate change. However, as pointed out in the 5 June 2012 issue of *Geophysical Research Letters*, the Eemian differed in essential details from modern climatic conditions. The Eemian is a poor analog for contemporary climate change, notably with respect to the rapid, ongoing disappearance of summer ice in the Arctic.

Even the conservative International Energy Agency (IEA) has thrown in the towel, concluding that "renewable" energy is not keeping up with the old, dirty standard sources. As a result, the IEA report dated 17 April 2013 indicates the development of low-carbon energy is progressing too slowly to limit global warming.

The Arctic isn't Vegas—what happens in the Arctic doesn't stay in the Arctic—it's the planet's air conditioner. In fact, as pointed out 10 June 2013 by research scientist Charles Miller of NASA's Jet Propulsion Laboratory, "Climate change is already happening in the Arctic, faster than its ecosystems can adapt. Looking at the Arctic is like looking at the canary in the coal mine for the entire Earth system." In addition, "average summer temperatures in the Canadian Arctic are now at the highest they've been for approaching 50,000 years" (and perhaps up to 120,000 years), according to a paper published online 23 October 2013 in *Geophysical Research Letters*. On the topic of rapidity of change, a paper in the August 2013 issue of *Ecology Letters* points out that rates of projected climate change dramatically exceed past rates of climatic niche evolution among vertebrate species. In other words, vertebrates cannot evolve or adapt rapidly enough to keep up with ongoing and projected changes in climate.

How critical is Arctic ice? Whereas nearly eighty calories are required to melt a gram of ice at 0 C, adding eighty calories to the same gram of water at 0 C increases its temperature to·80 C. Anthropogenic greenhouse gas emissions add more than two and a half trillion calories to Earth's surface every hour (about three watts per square meter, continuously).

Interactions among feedbacks are particularly obvious in the Arctic. For example, as reported in the 5 May 2014 issue of *Geophysical Research Letters*, "further reductions in seasonal ice cover in the future will result in larger waves, which in turn provide a mechanism to break up sea ice and accelerate ice retreat."

Ocean acidification associated with increased atmospheric carbon dioxide is proceeding at an unprecedented rate— the fastest in 300 million years—leading to great simplification of ecosystems, and capable of triggering mass extinction by itself. Already half of the Great Barrier Reef has died during the last three decades and the entire marine food web is threatened. As with many attributes, the Arctic Ocean leads the way in acidification. Similarly to the long lag in temperature relative to increase greenhouse gas emissions, changes in ocean acidity lag far behind alterations in atmospheric carbon dioxide, as reported in the 21 February 2014 issue of *Environmental Research Letters*.

An increasing number of scientists agree that warming of 4 to 6 C causes a dead planet. And they go on to say that we'll be there much sooner than most people realize. According to a paper in the 24 November 2013 issue of *Nature Climate Change*, warming of the planet will continue long after emissions cease. Several other academic scientists have concluded, in the refereed journal literature no less, that the political target of 2 C is essentially impossible (for example, see the review paper by Mark New and colleagues published in the 29 November 2010 issue of the *Philosophical Transactions of the Royal Society A*). The German Institute for International and Security Affairs concluded on 2 June 2013 that a 2 C rise in global average temperature is no longer feasible (and Spiegel agrees, finally, in their 7 June 2013 issue), while the ultra-conservative International Energy Agency concludes that, "coal will nearly overtake oil as the dominant energy source by 2017 . . . without a major shift away from coal, average global temperatures could rise by 6 degrees Celsius by 2050, leading to devastating climate change."

If these views seem extreme, consider (1) the 5 C rise in global average temperature fifty-five million years ago during a span of thirteen years, as reported in the 1 October 2013 issue of *Proceedings of the National Academy of Sciences*, and also (2) the reconstruction of regional and global temperature for the past 11,300 years published in *Science* in March 2013. One result from the latter paper is shown in the figure below.

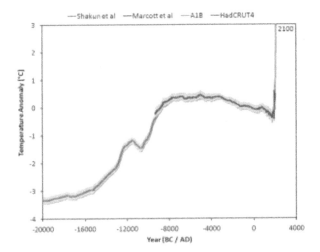

It's not merely scientists who know where we're going. The Pentagon is bracing for public dissent over climate and energy shocks, as reported by Nafeez Ahmed in the 14 June 2013 issue of the *Guardian*. According to Ahmed's article:

> *Top secret US National Security Agency (NSA) documents disclosed by the Guardian have shocked the world with revelations of a comprehensive US-based surveillance system with direct access to Facebook, Apple, Google, Microsoft and other tech giants. New Zealand court records suggest that data harvested by the NSA's Prism system has been fed into the Five Eyes intelligence alliance whose members also include the UK, Canada, Australia and New Zealand.*

In short, the "Pentagon knows that environmental, economic and other crises could provoke widespread public anger toward government and corporations" and is planning accordingly. Such "activity is linked to the last decade of US defense planning, which has been increasingly concerned by the risk of civil unrest at home triggered by catastrophic events linked to climate change, energy shocks or economic crisis—or all three." In their 2014 Quadrennial Defense Review, the US military concludes the following: "Climate change poses another significant challenge for the United States and the world at large. As greenhouse gas emissions increase, sea levels are rising, average global temperatures are increasing, and severe weather patterns are accelerating."

CHAPTER 3

HOSPICE OF THE HEART: WHY DO I FEEL MORE ALIVE THAN EVER?

BY CAROLYN BAKER

Do you think about yourself, and how you might profit or escape from a situation? Or do you think about others, and how you can help? Progress on the path, and a sign that you're well prepared for death, is when the former changes into the latter, when you default not into selfishness but into selflessness. If you're uncertain about what to do in a situation, just open your heart and love.

~Andrew Holecek, Preparing To Die~

The optimum preparation for death is a wholehearted opening to life, even in its subtlest turnings and changes.

~Stephen Levine, Healing Into Life And Death~

If you read and processed Guy's chapter on the "hard" science of climate change, you must be having some feelings about it. I suspect those could range from an initial response of "This can't be true, and I'm not going to continue reading" to heart-racing terror, or to sorrow, rage, or despair. But notice which responses cause you to feel your body—and which one doesn't.

We are all robustly defended against the very bad news regarding catastrophic climate change. We prefer to hear the soporifics of President Obama and the Environmental Protection Agency, which "reassure" us that the United States is going to take dramatic action to fight it. Go back to sleep, no worries, we're doing something about it. You breathe a sigh of "relief" and think that maybe your grandchildren will have a future. Maybe the human species will continue a bit longer—or a lot longer. Meanwhile, what is so easy to miss in Guy's twenty or so pages of climate science documentation is that the implications are so immediate, so momentous, that the real issue is not your grandchildren's future, but yours!

In the 1950s, Stanford University psychologist Leon Festinger said, "A man with a conviction is a hard man to change. Tell him you disagree and he turns away. Show him facts or figures and he questions your sources. Appeal to logic and he fails to see your point." Chris Mooney in his recent article "The Science of Why We Don't Believe in Science," opens with Festinger's quote, and then proceeds to illumine us about what we tend to do with facts when they don't fit our hopes and dreams, but more to the point, when they suggest that our lives are in imminent danger.

When we think we're reasoning, we may in fact, be rationalizing. Mooney says the following:

In other words, when we think we're reasoning, we may instead be rationalizing. Or to use an analogy offered by University of Virginia psychologist Jonathan Haidt: We may think we're being scientists, but we're actually being lawyers (PDF). Our "reasoning" is a means to a predetermined end—winning our "case"—and is shot through with biases. They include "confirmation bias," in which we give greater heed to evidence and arguments that bolster our beliefs, and "disconfirmation bias," in which we expend disproportionate energy trying to debunk or refute views and arguments that we find uncongenial.

When data are difficult to process, we tend to shoot the messenger. For example, we attack the personality of the messenger or find something about them to dislike so that we need not trouble ourselves with the message. We shrug off the message with "He's so strident," or "Of course she would say that, she's an old lady with one foot in the grave."

We would be hard pressed to find a culture in modernity that so adeptly and adroitly wards off anything it doesn't want to deal with as the culture of industrial civilization. One of the myriad reasons people contact me for life coaching is that they are struggling on a daily basis to hold in their bodies and minds the facts about abrupt climate change and the collapse of industrial civilization alongside the performance of their regular tasks and the maintenance of friendships and family relationships. They commonly report "feeling crazy" as a result of the cognitive dissonance they experience moment to moment as they attempt to navigate two different world views.

What my clients have already begun to understand before they contact me is that if one chooses to embark on the journey of truth, one will invariably experience a number of "disruptive dilemmas," as one of my clients names them, that will alter virtually everything in one's life, and from this journey, there is literally no return. Early in 2014 my article "Embarking on the Journey of Consciousness: Staying on the Train," names four distinct features of the process:

- When confronting a new piece of information about our planetary predicament, each of us chooses whether to ingest and assimilate the information or not. If we are kind to ourselves, we ingest a bit of it, allow it to distill, and then acquire more when we feel ready. Furthermore, genuine kindness to ourselves also means that we pay attention to the emotions that are stirred—fear, anger, grief, despair, and more. Rather than attempting to flee from uncomfortable feelings by engaging in intellectual debates about the accuracy of the information with which we are confronted, we notice our emotions even as we engage in deliberation.

- Early on our journey appears to be nothing more than a project of gathering information. As we progress, we may experience it as our principal survival tool. We may tell ourselves that because "knowledge is power," the more we know, and the more can protect ourselves and our loved ones.

• At some point in the journey, we move beyond simply gathering information, and rather than our owning the journey, it begins to own us. Invariably, whether we consent or not, we enter territory that I can only describe as "spiritual." In writing about spiritual journeys, my friend and colleague, Terry Chapman, defines it as "the ongoing, transformative experience of intentional, conscious engagement with what the sojourner perceives as the presence of divinity." Another word for divinity might be "the sacred," or "something greater or even "existential"—issues having to do with meaning and purpose. Typically people on such a journey choose to arrange their lives not so much around survival as around service. The core issue of one's life becomes not how long they can stay alive, but how they can contribute to the Earth community. One becomes infused with compassion and gratitude. No day, no being, no experience is ordinary, but rather, imbued with meaning.

• Whether we acknowledge it early or late in the journey, we eventually grasp that what we are ultimately confronting is our own death. The sooner we can honestly confront our mortality, allowing ourselves to actually feel it in the body, the easier it becomes to ingest and assimilate more distressing information. For example, when I have led some people in a "die before you die" exercise, they have often told me that once they sat with their own death and how it might actually feel, they felt more capacity to face not only near-term extinction but a variety of losses and catastrophes.

Two things passengers on the train of truth are compelled to confront are emotions and existential dilemmas. People who have previously spent most of their time inhabiting the mental realm discover that the body is persuasively pulling them farther away from that realm and more directly into the emotions and the heart. They frequently contact me because in a culture so disconnected from the heart they are attempting to manage their emotions with little or no awareness of how to do so. One of my interventions is assisting people in becoming familiar with fear, anger, grief, and despair. I also paradoxically explain that if they are willing to befriend these emotions, they are very likely to experience a deeper quality of joy and wellbeing. In fact, last year I published an article series entitled "What Collapse Feels Like, Parts 1-5" in which I outlined the process of working with the so-called "dark emotions" which many people tell me are challenging to feel, yet when they allow themselves to do so they feel much more alive than when they were obliviously ensconced in the "Coca-Coma" of industrially civilized consumerism.

One emotion in particular seems most salient and pervasive in the awakening process, namely grief. Often when people explore their fear, anger, and despair, they unearth a well of deep grief that seems to underlie many other emotions. The culture of civilization is a culture of profound loss because it has stripped from humans so much of what makes them human and has estranged our species from most other living beings on the planet. Collapse and abrupt climate change are presenting humans with the opportunity to discover and grieve the losses inflicted on the entire Earth community. Much of my work is, in fact, about assisting people in discovering the myriad opportunities the death of species and the death of our living arrangements, as well as our own personal demise, can offer. This is the essence of what author and environmental attorney Zhia Woodbury names "Planetary Hospice."

A HOSPICE PERSPECTIVE

Hospice is generally a specialized form of care for individuals who are very ill or are approaching death. For the most part, it is voluntary, and people choose to be admitted to hospice or decline. Many health insurance plans include hospice coverage while others do not, and few individuals without health insurance coverage have the option of professional hospice care. The word "hospice" along with the words "hospital," "hospitality," and "hospitable," are rooted in the Latin word "hospitium," which relates to providing temporary shelter and care for travelers. Thus on the journey from this life to whatever lies beyond it, hospice not only marks a transition but an extraordinary opportunity for the dying to receive uniquely humane and individualized care.

In this book, we assume that our planet is in a state of hospice, and we also affirm that a hospice attitude toward our predicament is the most appropriate response. What, then, is a hospice attitude?

As noted above, two things the journey of consciousness requires us to confront are emotions and existential dilemmas. In terms of emotions, hospice is a place where we intentionally focus on the emotions that arise as a result of our willingness to engage with the reality of collapse and abrupt climate change. In hospice, grief often eclipses all other emotion, as it should, given that in contemplating death, one is often besieged with the totality of losses throughout one's life—lost relationships, lost opportunities, regrets that cannot be repaired, situations which John James and Russell Friedman describe in The *Grief Recovery Handbook* as those that we wish had been "different, better, more." Moreover, it is one thing to contemplate the losses of one's life and yet another to confront the loss of our planet and most or all species on it.

I believe that grief is the doorway not only to discovering the richness and reward of savoring all of our emotions but that it is also the doorway to engaging with the spiritual dilemmas of meaning and purpose. In his powerful book on grief, *Entering the Healing Ground: Grief, Ritual, and the Soul of the World*, Francis Weller states that "Opening to our sorrow connects us with everyone, everywhere. It is sacred work." Not only does grieving connect us with people but with the Earth itself because to be of the Earth is to grieve.

Some would argue that grieving is a passive activity that serves no one except the griever, but as Francis Weller states above, it is "a powerful form of soul activism." In fact, we may want to consider the possibility that the Earth longs for our grief in the face of what is being inflicted upon it. I believe that we must feel remorse and regret for our participation and collusion in its exploitation, but that we must also consciously grieve these.

Grief work should not be undertaken without support. While many people contact me for that specific purpose, we can also support each other as friends or in small groups as we bear witness to and honor each other's grieving process.

The German poet Rilke understood the healing power of grief and also the capacity of grief to transport us to the sacred within us. Allowing himself to descend into the dark shadows of grief, he discovered that "Deep in the darkness is God," and that "A great presence is stirring inside me; I believe in the night." Rilke uses the word "God," but I do not allow it to become a barrier between me and what I call "something greater" that lives in my core. When we consciously grieve, we taste the mystery and awe that permeates our humanity and awakens us to our intimate connection with each other and the more-than-human world. Moreover, our grief usually results in a deeper capacity to feel joy, appreciate beauty, and savor humor as clarified in all of my writings.

Some individuals are so profoundly distressed by the demise of the planet that they have become engulfed by rage toward the human species, and they chant an incessant litany of "the Earth would be so much better off without humans." While I understand this perspective, I have not found it useful—for the individual or the Earth community. In my 2014 article "Mad Hominem: Why Hatred of the Human Species Is Not Helpful," I noted that rather than becoming consumed with contempt for our species, it is much more useful and important to fully engage in remorse for our part in the eco-side and also recognize that beneath our contempt is probably deep grief. Holding a vice grip on our anger guarantees that we will not feel the sorrow of our predicament, and our eyes can remain dry and our bodies armored. As a result, we betray all that is good and decent in our humanity, and doing so makes not one iota of a difference to other suffering species.

Rage also prevents us from being as kind as we could be to all living beings. Unless we weep repeatedly, we will not know or practice the kindness that is necessary at this time of planetary hospice. Hospice is a time for compassion—for ourselves and all beings on Earth. Collective salvation may never come, but if any personal salvation is to come, it will only come as author Charles Eisenstein says: "When we face up to the ugliness of our own past and feel the mirror image of the pain of every slave lashed, every many lynched, every child humiliated. One way or another, we must weep for all of this." We must weep for the more-than-human world victimized by our madness—and we must weep for ourselves.

But what else do people do in hospice? Do they simply sit around crying all day? Indeed if they are well enough, they can serve their fellow hospice dwellers by doing acts of kindness and service. They have ample hours to reflect on their lives and perhaps make amends with people (and even members of the more-than-human world) they have offended. People in hospice might also have the opportunity to enjoy singing together, sharing food, or just simply sitting by the bed of another person and listening to their story. Some people have reported that their time in hospice was surprisingly filled with meaning, merriment, and even joy. "Only paradox comes close to comprehending the fullness of life," said Carl Jung. When we experience moments of joy, grief is not far away because they are inextricably connected, and when we allow ourselves to grieve, joy is waiting in the wings to bless us, as echoed in these words attributed to William Blake: "The deeper the sorrow, the greater the joy."

WE ARE THE UNIVERSE: DEATH AS DOORWAY

Thomas Berry was born in Greensboro, North Carolina, in 1914 and entered a Passionist Order monastery at the age of twenty. He was magnetically drawn to the writings of Jesuit priest and philosopher Teilhard de Chardin who was a paleontologist and deeply engaged in what today we would name "eco-spirituality." Many years after receiving a Ph.D. in history,

Berry became profoundly enchanted with the Earth and ceased calling himself a theologian in favor of the term "geo-logian." Soon Berry began writing about "the Great Work," which he defined as the re-sacralizing of the universe and the journey of humans to understand that they are not separate from the universe but rather that they are the universe. The Great Work, according to Berry, is to come to know ourselves as planetary humans—as beings intimately connected with the universe in order to enhance the flourishing of the Earth community.

Linda Buzzell and Craig Chalquist in their recent book, *Ecotherapy: Healing with Nature in Mind*, offer a clear description of Berry's Great Work:

> *What is unique to the Great Work of our time is a dawning realization, largely through the discoveries of science, that we live in an evolving universe with an irreversible sequence of emergence, not simply one based on the cycles of the seasons. Thus we require the healing power of story to place us in that emergence in a meaningful way and to give us a sense of its import. Many call this 'New Story' the source for a new faith in the human prospect. Its unifying narrative points to our awakening to meaning and wonder in a cosmos that has become increasingly impersonal under the sway of Western science and the classical humanist tradition. Moreover, knowledge of our origins creates a sense of continuity with the past and with the future, instilling in us a sense of concern for future generations. Understanding the evolutionary past is*

not just a scientific exercise, but an exercise in identity. In this way, we come to understand the relevance of the Great Work to our generation. (271-272)

Whether or not one grasps the likelihood of near-term human extinction, the supreme relevance of understanding and engaging with the New Story is not so much about our longevity, but rather our identity, even if our immersion in it is new and very short-lived.

Before his death in 2009, Berry worked closely with physicist and mathematical cosmologist Professor Brian Swimme at California Institute of Integral Studies, regarding a new approach to science and history that emphasizes humanity's connection with the universe. Influenced by his collaboration with Berry, Swimme produced a series of video lectures entitled "The Powers of the Universe" in 2004 that articulates not only the fundamental functions of the universe but also how humans can consciously partner with them. Berry and Swimme's assumption is that our psychic identity as humans is too limited for the power we have taken on the Earth. Our influence stretches around the Earth, yet our allegiance is too small—for example, our allegiance to family, country, and corporation. As a result, we've lived and practiced domination and human centered-ness that has caused degradation and destruction.

The term "powers of the universe" originated in an essay Berry wrote entitled "The World of Wonder," included in the 2014 book *Spiritual Ecology: The Cry of the Earth* that was edited by Llewellyn Vaughan-Lee. Berry emphasizes that native people around the world have recognized these powers for millennia and have utilized them as a reference point in human affairs. Thus, Swimme's comprehensive analysis of the powers in scientific terms is merely a cataloging of ancient indigenous knowledge, delivered systematically for the rational, linear Western intellect.

The essential focus of the series is to assist the viewer in becoming a student of all living beings on Earth and build mutually enhancing relationships with them that work for all life. Ann Amberg, a student of Swimme and Berry's work, has additionally incorporated into her articulation of the Powers of the Universe, the work of Riane Eisler whose focus is partnership in relation to the universe, the human community, and individual relationships.

It is important to understand that even in the face of abrupt climate change and near-term human extinction, we can engage with the Powers of the Universe, not with the intention of "saving" the planet, but rather as a procedure for partnering with the universe in concert with the likely cessation of all life on Earth. Extinction may prevail, but as with the hospice juncture we may be, as Brian Swimme states, "released into the essence of who we really are." Additionally, our willingness to partner with the universe facilitates our commitment to making the demise of other species as painless as possible. For me, the Powers of the Universe offer a doorway to the essence of the hospice perspective while at the same time providing one of the most inspiring and energizing models for living life with exuberance and purpose.

THE POWERS OF THE UNIVERSE

1) *Centration* is how the universe centers on itself. As it billows out, it fractures into a multiplicity of centers—for example, galaxies. Everything in the universe begins and moves out from a center. Life could have remained at a cellular level, but it pushed forward into multi-cellular beings. We cooperate with centration by removing ourselves from the values of modern industrial consciousness and developing deep, intimate connections with nature. That's to say that we make nature, not the values of the human ego, our fixed point in a changing universe.

2) *Allurement* is the power of attraction or the gravitational force that holds the universe together. It is as if it was not enough for the sun and the Earth to be bonded for several billion years. The bond evolved into the deeper intimacy of photosynthesis. We most effectively cooperate with allurement by allowing ourselves to be drawn to all that is wild in nature. We do not make allurement happen. Rather it

happens to us, and we surrender or resist. By surrendering to allurement, we naturally move into a pursuit of beauty.

3) *Emergence* is exemplified in moments of birth in the universe such as when atoms are formed, stars and galaxies come forth, and various forms of advanced life show up. Emergence also takes place at the end of one era and the birth of another. Even in the most horrific moments of suffering, emergence is possible. For emergence to happen, order has to be destroyed. The emergent human is one who is likely to be disturbing to the forces of power and the structures of industrial consciousness. He/she understands that they exist to participate in mutually enhancing community.

4) *Homeostasis* is an ongoing function of the universe in the desire of systems to protect and maintain their integrity. Humans have dramatically altered the homeostasis of a plethora of natural systems, but we can support homeostasis by restricting human activity that would disturb it. Yet another aspect of homeostasis in a time of extinction is simply telling the story of Earth's emergence and evolution to our children and to each other. Telling the story reinforces our awareness of the rights of all beings to

inhabit the Earth community. Many indigenous cultures have creation stories that they tell and re-tell to each other on a regular basis. In the last days of Thomas Berry's life, he was asked what final words he has for humans. Reportedly, he replied, "Tell the story."

5) *Cataclysm* is synonymous with the realities of destruction, degradation, and collapse. Every one hundred million years there has been a mass extinction, and we are presently living in the midst of one, hence the second law of thermodynamics: things fall apart and break down. Life in the universe constantly rises and falls. As Brian Swimme notes, "Cataclysm is happening. The choice before us is whether we will participate consciously." As we consciously partner with cataclysm, we invariably surrender those parts of ourselves that have contributed to ecocide and extinction. We allow them to be torn apart and shredded. According to Ann Amberg, abrupt climate change and the collapse of industrial civilization represent opportunities for humans to partner with cataclysm. This is the crux of the hospice perspective. Call it "cataclysm," call it "transformation," or call it as does Joanna Macy: "the greening of the self."

6) *Synergy* describes working and creating together—the formation of mutually enhancing relationships. One example in nature is the reality that many plants cannot exist without pollinators. Yet another is photosynthesis and the sharing of information by cells. In industrial society, we are shaped to be anti-planetary. Everything about the Earth is "other," and all living beings have been "othered" by industrial consciousness. Synergy really means that even in a time of extinction, we arrange our lives and our thinking around the principle that our core identity is Earth, and we form the most mutually enhancing relationships possible. Indeed it may be that one of the final tasks of humanity in the wake of near-term extinction is to discover at last that we have a relationship with every person, thing, and event in our lives—a concept on which I elaborate in detail in my forthcoming 2015 book, *Love in the Age of Ecological Apocalypse: The Relationships We Need to Thrive.*

7) *Transmutation* may be appreciated in the way that the universe forces itself out of one era and into another. It doesn't just create new forms of life, but it makes the conditions of former life impossible. We are in the midst of such an era. Our job is to participate, and we do so by focusing with great

discipline on living our lives in intimate relationship with the Earth and its needs. The Earth becomes our *raison d'etre* and the guiding principle of our behavior and relationships. Certain aspects of us must be relinquished. The industrially civilized ego must be diminished so that the wild, instinctual, animal self may partner with our more-than-human relatives.

8) *Transformation* relates directly to transmutation. Whereas transmutation is the power of change at the individual level, transformation is about the change worked into the whole through individual efforts. According to Swimme, one of the great shocks for science was that the universe arrived at life so quickly. We now understand that life is central to the universe, and physicist Freeman Dyson notes, "life may have a larger role to play than we have imagined." In order to arrive at life, the universe had to transform itself repeatedly. In the universe, a particular species for survival purposes may move away from a value deeply embedded in its DNA, and a genetic mutation eventually occurs. The mutation is folded into the DNA over time, and the species is better able to thrive as a result. For humans, transformation is about putting forth a value that can shift all of life. Gandhi and Martin Luther King Jr.

are but two examples. Even in an era of extinction, we can offer new ideas and new ways of living that create a shift in the lives of those who embrace them.

9) *Inter-Relatedness* is the capacity to identify worth and needs. A synonym for this is "care." Parental care was invented in the universe, and it made survival more possible. In fact, natural selection favored species that valued it, and the power of care deepened when mammals evolved. Within industrial society, the paradigm is a "use" relationship with other beings. We assume that they are here for our use, and use-relatedness results in the assumption that life has no meaning other than what we can acquire through use. Conversely, imagination allows us to find meaning, and the vehicle through which we usually discover meaning is care. Obviously, we cannot use other members of the Earth community if we genuinely care for them.

10) *Radiance* allows us to consciously respond at a feeling level to the magnificence of the Earth and the radiance of other beings. Far deeper than human language is the radiance the universe is displaying at all times. Everything in the universe radiates light, or as Swimme suggests, "This is the primary manner in

which the universe communicates with itself." It is also the primary manner in which the universe communicates with us, but we cannot receive the communication unless we listen and look. When we fall deeply in love with the Earth community, we receive and reciprocate radiance in all of our relationships.

As a result of utilizing Swimme's model of the *Ten Powers Of The Universe* in recent years, I have experienced an unprecedented reverence for and relatedness with the universe. This synopsis of the *Ten Powers* is unfairly brief, and therefore, I highly recommend further engagement with Swimme's DVD or online courses on the *Ten Powers* offered by Ann Amberg. Nevertheless, I have included the summary as an invitation to explore the *Ten Powers* and utilize them as one paradigm for inhabiting hospice. They call us to both rest in our demise and live with compassion, presence, and extraordinary exuberance going forward.

FROM EGO-CENTRISM TO ECO-CENTRISM

The Ten Powers of the Universe is but one model for consciously deepening one's connection with the Earth. They exemplify a long-overdue trend in human psychology, namely ecopsychology, which, though novel to modern inhabitants of industrial civilization, is as old as our species. Ecopsychology is essentially the synthesis of the psychological and the ecological that allows the psyche to bond with the Earth for the purposes of healing both the self and the Earth. It is based on the theory of biophilia introduced by Harvard zoologist E.O. Wilson, which he defined as an innate emotional affiliation humans have with other organisms. Whereas a typical session of psychotherapy might occur in a consulting room where the therapist asks the client about their relationship with their mother, a session of ecotherapy might take place outdoors where the therapist inquires about and observes the client's relationship with "Mother Earth."

In *Ecotherapy: Healing with Nature in Mind*, Linda Buzzell and Craig Chalquist note that "During this global environmental crisis, it is crucial for mental health clinicians to begin to understand the connection between the epidemics of mental distress in modern industrial societies and the devastating impact of the destruction of our own habitat." In fact, all of humanity has colluded in some manner with this eco-side, and we carry within us tremendous grief, guilt, emptiness, and fear. Ecopsychology assists us in recognizing our part in destroying the Earth and doing what we can to heal it.. Buzzell and Chalquist further explain by stating the following:

> Most therapy clients don't realize that much of the grief, shame, emptiness, and fear they struggle with may be a natural response to the unnatural way we live. The loss and death of so many living beings, guilt over our individual and collective complicity in

these deaths, and the ongoing distress of Earth, air,
and ocean life all around us in our very bodies are all
sources of stress. (47)

We frequently attribute mass extinction and eco-side to our disconnection from nature, but how might that estrangement be healed? Do we simply wander in nature until we feel something? Although it may sound simplistic, my partial answer is yes! On the other hand, it is important to know why we are wandering and what we are seeing, hearing, feeling, smelling, or tasting as we do so. Ecopsychology provides guidelines for this kind of exploration.

In his book *Nature and the Human Soul*, psychologist and author Bill Plotkin, explains, "Every human being has a unique and mystical relationship to the wild world, and . . . the conscious discovery and cultivation of that relationship is at the core of true adulthood." Note that Plotkin is not only saying that an intimate relationship with the universe is important, but that it is "the core of true adulthood." No wonder a culture torn asunder from nature is emotionally a culture not of adults but of adolescents at best and children at worst.

Derrick Jensen notes that *Nature and the Human Soul* "provides a road map to help us remember how to be human— which means how to be a human being in relationship to the natural world, to our home." Moreover, the book introduces an ecopsychological template of human development rooted in the cycles and qualities of the natural world. It takes the reader through all of the stages of human development specifically in relation to the Earth and its functions and suggests particular tasks that confront us at each stage, which if we are willing to embrace will facilitate our becoming wise elders at any age—mature men and women whose fundamental task above all else is partnering with and protecting the Earth.

From my perspective, it does not matter when one begins reading and practicing the tools in *Nature and the Human Soul*. For both the millennial and the senior citizen, Plotikin's work is a stunning manual for conscious living even in the face of conscious dying. In the final pages of the book, Plotkin notes that for the wise sage woman or man, "the prospect and process of dying is experienced as a natural and joyful return to spirit, a merging with the Mystery from which we sprang."

THE MANY REVITALIZING USES OF DEATH

I am neither a hospice nurse nor a hospice volunteer, but I have been present with the dying and, as stated above, have lived through two cancer diagnoses. In *Preparing to Die*, Andrew Holecek, a Tibetan Buddhist teacher, states, "If we acknowledge death and use it as an advisor . . . it will prioritize our life." The author further states that one of the most important preparations for death is to lead a genuinely spiritual life in which love is the key. Additionally, a spiritual life is a life of acceptance. This is not to say that when we witness injustice or eco-side, we simply ignore it or bypass it spiritually and emotionally by meditating or using some other technique to achieve "inner peace." Rather, cultivating a perspective of love and compassion in the face of near-term human extinction is a daunting but imperative practice.

WHEN SURRENDER MEANS NOT GIVING UP

In April 2014, following a conversation with my friend, Andrew Harvey, author of *The Hope: A Guide to Sacred Activism* and founder of the Institute for Sacred Activism, I penned an article proposing a "New Sacred Activism" that vigorously pursues protest and an activist agenda, but at the same time recognizes the power of acceptance.

I've recently noticed some longtime activist voices verbalizing a new perspective and one that some would label as "defeatist." Writing in his blog titled "How to Save the World," Dave Pollard offered a piece called "In Defense of Inaction," in which he states the following:

> *No one is in control. The enemy, if there is one, is not a cabal of elites, but a set of co-dependent collapsing systems that every one of us has a vested interest in trying (insanely) to perpetuate. Systems we have all helped co-create and are almost all dependent on . . . The question we must each ask ourselves, I think, is this: If we acknowledge that our systems and hence our civilization cannot be reformed or 'saved,' what can we do now that will make a real difference, for the future, in our communities and for those we love? . . . The insanely rational answer to this*

question, I think, is (a) probably nothing, and (b) it's too early to know.

Subsequently, Wen Stephenson's "Let This Be the Last Earth Day," in *Nation Magazine* pleads:

> *End the dishonesty, the deception. Stop lying to yourselves, and to your children. Stop pretending that the crisis can be "solved," that the planet can be "saved," that business more-or-less as usual—what progressives and environmentalists have been doing for forty-odd years and more—is morally or intellectually tenable. Let go of the pretense that "environmentalism" as we know it—virtuous green consumerism, affluent low-carbon localism, head-in-the-sand conservationism, feel-good greenwashed capitalism—comes anywhere near the radical response our situation requires... The question is not whether we're going to "stop" global warming, or "solve" the climate crisis; it is whether humanity will act quickly and decisively enough now to save civilization itself—in any form worth saving. Whether any kind of stable, humane and just future—any kind of just society—is still possible.*

Isn't this just giving up or giving in—the pathetic whining of defeatists? How can any respectable activist utter these words?

I argue that in fact, the perspectives articulated by Pollard and by Stephenson are utterances of concession and of extraordinary courage. The word I choose to describe their perspective is "surrender." But isn't "surrender" synonymous with "giving up"? Aren't Pollard and Stephenson really suggesting that we abandon the struggle, acquiesce, go back to business as usual, eat, drink, be merry, or possibly take our own lives?

In my recent article, "What Does It Mean to 'Do Something' about Climate Change?" I noted the following:

> 'Doing something' implies that developing nations of the world and the fossil fuel industry will come together and: 1) Agree that climate change is actually happening; 2) Understand that the situation is so dire that humanity's living arrangements must be radically altered; 3) Sacrifice their economic security and industrial profits to significantly reduce carbon emissions; 4) Agree to the reality of climate change and the altering of their living arrangements in time to prevent another 2 degree C rise in temperature.

I then asked the reader what they genuinely, realistically think can be done about catastrophic climate change, particularly in the face of more than thirty self-reinforcing feedback loops that are proving the process unstoppable and irreversible.

I believe that what both Pollard and Stephenson may be echoing is a third perspective that Andrew Harvey and I are naming "The New Sacred Activism." That is to say that the fundamental issue is that we are being challenged to move beyond the triumphalist assumption that we can and must, through our activism, defeat capitalism, catastrophic climate change, economic corruption and collapse, and, yes, the extinction of species, including our own. Our current predicament indeed compels us to transcend the binary inference that if we do not conquer a diabolical system, we are only colluding with it and exhibiting shameful cowardice.

Contrary to our cherished assumption of vanquishing all forms of injustice, we must ask ourselves if we are willing to put love into action even if we don't physically survive. The extremity of the crisis does not limit Sacred Activism but expands it because we make ourselves available to 1) bearing witness to the likely irreversible horrors of climate chaos and 2) commitment to compassionate service to all living beings who suffer with us. This requires unwavering engagement with serving the Earth community and practicing good manners toward all species in order to make their demise, and ours, easier. Taking one's own life or succumbing to escapist self-medication is easy. Commitment to a life of service and fortifying one's own connection with the sacred, thus deepening one's sense of meaning and purpose, constitute a far more daunting and painful path.

Very often people who receive a terminal medical diagnosis report that while the announcement was heartbreaking, terrifying, and profoundly unfair, they experienced a certain kind of liberation in the process. Learning that they had only a limited time to live altered every aspect of their lives, particularly their quality of life, their decisions about how they wanted to engage with the remainder of their days, and their relationships with everyone and everything. Similarly, the reward of accepting the reality of near-term extinction is liberating even at the same time that it is agonizing. Something far more meaningful and momentous beyond our own physical survival becomes available to us.

THE HERO ARCHETYPE

Human consciousness is deeply influenced by personal and cultural archetypes or universal themes of which we may or may not be aware. A few examples of archetypes are mother, father, hero, savior, martyr, warrior, maiden, crone, healer, and many more. If we are not aware of the archetypes that influence us, we may unconsciously live them out in both creative and destructive ways.

Mythologist Joseph Campbell wrote and taught more about the hero archetype than perhaps anyone in modern times. Campbell simply defined the hero/heroine as "someone who has given his or her life to something bigger than oneself." According to Campbell, the hero's task is threefold: separation, initiation, and return. He/she usually experiences some sort of unusual circumstances at birth, sustains a of traumatic wound, acquires a special weapon that only he/she can use, and proves him/herself by way of some sort of quest or journey through which he/she is forever changed. The hero's journey is one of death, rebirth, and transformation on which he/she embarks for the wellbeing of the community. Meanwhile, it is the hero, as well as the community, who is transformed.

The pitfalls along the journey are many, but one of the most common and also the most injurious to the hero and the community is to become inflated with one's heroic mission. One's passionate commitment to the journey makes avoidance of this particular snare exceedingly difficult. Along the way, the hero may find him/herself becoming inflated but can always choose to pause and reflect upon the purpose of the journey and the "something bigger" that compelled the hero to begin the journey in the first place. This is an opportunity to surrender to the ultimate purpose of the journey and the spiritual forces that motivate and support the hero. In all of mythology, the ultimate purpose is the transformation of conscious, both one's own and that of the community. In mythology, failure to surrender to the larger purpose of the journey guarantees the hero's demise.

In Greek mythology, the hero was always aware of the seduction of thinking himself equal to or wiser than the gods. Whenever heroes succumbed to this temptation, they began plummeting toward their demise. Perhaps the most famous example of the inflated hero is Icarus who, determined to soar, flew too close to the sun whereby his wings that were made of wax began melting, and he fell into the sea. Whether the wax wings of Icarus or the vulnerable heel of Achilles, the fundamental lesson with which all Greek mythological heroes were confronted was their human limitation and the consequences of forgetting those.

The hero symbolizes courage and sometimes appears as a warrior as well, but whether hero or warrior, courage is one of his/her stellar characteristics. One lesson the hero/warrior must learn is when to fight relentlessly and when to exercise restraint or surrender, for whatever reason, to the dilemma with which he/she is confronted. The Shambhala Warrior, for example, receives training in fighting spiritual rather than physical battles. Discernment regarding restraint or full-on combat is pivotal, and at all times the warrior must examine the heart and will. He/she does not continue fighting at all costs simply because "that is what warriors do." Rather, the spiritual warrior considers whether or not it is time to stop fighting altogether or whether it may be time to change strategy. Fighting may not mean "winning" in the heroic sense, but fighting for reasons that surpass even one's survival.

Bill Plotkin writes, "The mature hero endures a descent to the underworld, undergoes a decisive defeat of the adolescent personality (a psychospiritual death or dismemberment), receives a revelation of his true place in the world, and returns humbly to his people, prepared to be of service according to his vision. This is equally true of the mature heroine."

As we confront catastrophic climate change that is likely to result in near-term human extinction, we must ask if we are willing to put love into action, even if we don't survive. Can we move beyond a triumphalist agenda? Accepting the possibility of near-term extinction is an agony, but an agony that liberates the spiritual warrior in the powers of truth and love in order to discover the diamond hidden in the darkness that cannot be discovered in relentless fighting in order to "overcome." The diamond can only be acquired by surrendering the need for anyone or anything to survive, even oneself. In the words of Andrew Harvey, this is "a glorious and terrible adventure, but it is the antidote to despair."

What we need now is not heroic victory but again, in Andrew's words, an "astringent maturity"—an entirely new level of adulthood that acts in ways that bring forth optimum joy, optimum healing, and optimum beauty which will leave seeds for whatever life might remain as most species on the planet face their demise. The sacred inspiration we require results not from false hope or finding solutions but from a state of active being in which we voluntarily enroll in radical psychological and spiritual training. If we haven't registered for this psychospiritual apprenticeship, then we will persevere in our triumphalist agenda and inadvertently perpetuate despair.

GRIEF AND THE NEW SACRED ACTIVISM

Of paramount importance in the new Sacred Activism is regular, conscious grief work. Unless activists mourn, they can easily be consumed with the fires of passion because their psyches are not tempered with the waters of grief. Conscious grieving is an integral aspect of the "astringent maturity" we develop as we balance hero/warrior courage with discerning acceptance of our predicament. In *Entering The Healing Ground: Grief, Ritual, And The Soul Of The World*, Francis Weller states:

> *Grief is the work of mature men and women. It is our responsibility to be available to this emotion and offer it back to our struggling world. The gift of grief is the affirmation of life and of our intimacy with the world. It is risky to stay open and vulnerable in a culture increasingly dedicated to death, but without our willingness to stand witness through the power of our grief, we will not be able to stem the hemorrhaging of our communities, the senseless destruction of ecologies or the basic tyranny of monotonous existence...Grief is...a powerful form of soul activism. If we refuse or neglect the responsibility for drinking the tears of the world, her losses and deaths cease to be registered by the ones meant to be the receptors of that information. It is our job to feel the losses and mourn them. It is our job to openly grieve for the loss of wetlands, the destruction of forest systems, the decay of whale populations, the erosion of soil, and on and on. We know the litany of loss, but we have collectively neglected our emotional response to this emptying of our world. We need to see and participate in grief rituals in every part of this country.*

For hundreds of years, members of the Dagara tribe of West Africa have practiced regular grief rituals because they believe that both the Earth and the community needs periodic releases of grief. Without doing so, they say, the heart becomes and remains hard, and this becomes toxic for the community. Always a community versus a private ritual, the Dagara experience that grieving together solidifies the community and makes conflict resolution less problematic and complicated. As a result of the community grieving together, members of the tribe also experience something that Westerners might not expect, namely a deepening of joy. Francis Weller shares his conversation with a Dagara woman immediately following a grief ritual that he attended. The woman displayed a radiant smile and seemed to exude joy from every pore. When he asked her how she could be so happy after engaging in a grief ritual, she replied, "I'm so happy because I cry all the time." Her response, it seems, echoes the profound words of Blake: "The deeper the sorrow, the greater the joy."

THE PERSONAL PERILS OF HEROIC ACTIVISM

In the spring of 2014, I was personally shaken by the death of my longtime activist colleague and friend, Mike Ruppert. After decades of political struggle and physical health challenges, Mike took his own life on April 13, 2014. Literally thousands of people attribute their awakening to our planetary predicament to Mike's efforts. The legacy he left for us is enormous, yet everyone close to Mike witnessed his reckless abuse of the body and the emotional wounding that ravaged his psyche and finally led to his suicide. As with many activists, Mike was only capable of surrender by terminating his physical life. While on the one hand he had every right to choose that path, I believe that his demise is a cautionary tale against heroic activism.

Heroic activism, particularly activism in which we do not grieve invariably, leads to burnout and compromised bodies and psyches. Rather than being self-indulgent, self-care, including grieving, is a spiritual practice that honors the corporeal container permeated by the sacred for the purpose of advancing its work.

LEAN INTO THE ANTHROPOCENE

At the website Welcome To The Anthropocene, this relatively new concept is explained:

Every living thing affects its surroundings. But humanity is now influencing every aspect of the Earth on a scale akin to the great forces of nature.

There are now so many of us, using so many resources, that we're disrupting the grand cycles of biology, chemistry, and geology by which elements like carbon and nitrogen circulate between land, sea, and atmosphere. We're changing the way water moves around the globe as never before. Almost all the planet's ecosystems bear the marks of our presence.

Our species' whole recorded history has taken place in the geological period called the Holocene—the brief interval stretching back 10,000 years. But our collective actions have brought us into uncharted territory. A growing number of scientists think we've entered a new geological epoch that needs a new name—the Anthropocene.

At the very least, humans are changing the chemistry of the planet. At worst, we are rendering it uninhabitable.

At first blush, it may seem that the extraordinary quality of intimacy with the Earth of which I have spoken would compel us to fight to our last breath to save it. When the heart is entwined with another living being, is not surrender to the demise of the beloved out of the question? We indeed stand in awe of Earth warriors such as Julia Butterfly Hill, Edward Abbey, and Derrick Jensen who have devoted their lives to defending and protecting our planet. Yet as with a terminally ill beloved human, we must now realistically assess our capacity to spare our own species and thousands of others from near-term extinction. The heroic option of martyrdom for the Earth is indeed a viable one. So also is a conscious consent to admit ourselves to planetary hospice as Zhia Woodbury so beautifully and brilliantly articulates in the 2014 article "Planetary Hospice: Rebirthing Planet Earth."

Woodbury, an environmental attorney and a practicing Buddhist, states that "The Great Anthropocentric Extinction is upon us. . . . Sober consideration of the current, cascading evidence leads to the inescapable conclusion that life as we have come to know it is, quite simply, at an end." At the conclusion of several pages of analysis of humanity's unwillingness to meaningfully address and attempt to reverse climate change, Woodbury concludes, "Our situation is regrettably, terminal."

On the one hand, the mental health profession should have a significant role in preparing us to face our demise, but it is ill equipped to do so. Eco-psychology, however, "is reinventing psychology by including 'the psychological processes that tie us to the world or separate us from it' in a more holistic vision of the human psyche (Buzzell & Chalquist, 2009, p. 17) that views humans and the world we inhabit as inextricably bound together....So from an eco-psychological viewpoint, the question now becomes what is the role of mental health professionals in preparing society for the end of life as we know it?"

The hospice model can be applied with the perspective that the coming catastrophe does not have to result in widespread fear, panic, dread, or hostility. As I have noted in a number of articles, such as "Preparing for Near-Term Extinction," "Fukushima and Catastrophic Climate Change: The Earth Community in Hospice," and "Hospice is a Busy Place," many people in hospice report that it was the most meaningful time of their lives. For them, it provided a sacred space in which to reflect deeply upon their lives—to evaluate relationships that were enriching, to make amends and restitution with respect to some relationships that were difficult and painful, to provide service to others in their hospice environment, and to prepare mindfully and reverently for death.

In this regard, Woodbury concludes, "If we are able to apply the same principles at a societal scale, then ecopsychologists and planetary thanatologists can become the kinds of spiritual midwives that will be needed to transform the planetary death/rebirth process from a painful dislocation rife with suffering and regret into a healing process for both the human race and the Earth itself—even into a Great Awakening."

Woodbury also applies Kubler-Ross's Five Stages of Grief to The Great Dying/Great Awakening. In fact, the author reminds us we are facing not only the possibility of near-term human extinction in the larger context of our planetary predicament, but also a plethora of endings on a smaller scale in terms of the limits of economic growth, energy depletion, and the end of life as we have known it in a variety of venues. Whatever our ultimate fate as a species, the fundamental assumption that progress is an infinite phenomenon to which humans are exceptionally entitled is unraveling at a dizzying speed as human consumption is now literally consuming life on Earth. Our spiritually puerile affirmations such as "Every day and in every way, everything is getting better and better," must be supplanted, not by "Every day and in every way, everything is getting worse and worse," but by a seasoned spiritual equanimity that wisely surrenders to whatever appears at the door of experience—a perspective so beautifully articulated in this excerpt from Rumi's marvelous "Guest House" poem:

This being human is a guest house.
Every morning a new arrival.

A joy, a depression, a meanness,
some momentary awareness comes
as an unexpected visitor.

Be grateful for whatever comes.
because each has been sent
as a guide from beyond.

The Great Dying cannot be practiced in isolation. Never before have humans required loving community to the extent that we do now. Woodbury notes the following:

> *Planetary hospice workers will be linked by a transmission of intention—the intention to be spiritual midwives for the rebirth of planet Earth. Just as the hospice movement today relies heavily on trained volunteers, so will planetary hospice rely heavily on the efforts of any and all who share the vision of the Great Dying as the dark night of collective humanity's soul, and who are equally committed to ushering the human race through this difficult 'night sea journey'—during which 'the sun sinks into the sea only to be devoured by the water monster—into the dawning light of a new day' (Washburn, The Ego and The Dynamic Ground, 1995, p. 21).*

Giving up, becoming defeatist? Is this the essence of the New Sacred Activism? Indeed it is not. As Woodbury notes, we must open to the "dark night of collective humanity's soul" and live as if every act, every task performed in daily life, every kindness expressed to another being and to oneself might be the last. This is one way I stay connected with the light in dark times. Walking in reverence, living contemplatively with gratitude, generosity, compassion, service, and an open heart that is willing to be broken over and over again. I do not always live the way I want to live. It's a practice, and practice never makes perfect. Practice only makes practice, and if I think it's perfect, I'm not practicing. Nevertheless, I'd rather stumble in the dark, finding an occasional candle to light the way, than become blinded by incandescent heroism. And so in this time of unprecedented darkness, find the light whenever possible, but most importantly be the light for someone else who may not be as familiar with the darkness as you are—and be willing to admit yourself to planetary hospice at the same time that you commit to being a hospice worker for the Earth community. That may be why you came here, and that is the New Sacred Activism.

WHAT'S TO LOVE ABOUT HUMANITY?

You may be one of the individuals to whom I alluded to above who holds a strong hatred of humanity and believes that the human species should be eradicated as soon as possible. If so, it will not be possible for you to willingly admit yourself to the hospice unit of near-term human extinction. The purpose of hospice is to prepare for death, not only in terms of getting one's logistical affairs in order, but softening and opening one's heart to an exquisite quality of living in the last hours of one's life. Hatred of humanity is unequivocally a defense—a warding off of feeling the hideous, catastrophic, gut-wrenching consequences of humanity's destruction of the planet. You certainly have the option to despise humanity, which raises the question of why you are still hanging around. You also have another option: embrace the finality of near-term extinction with openness to feeling all of the feelings that doing so will invariably evoke.

The real question here is not what's to love about humanity, but what's to love about you. If your hatred results from remorse, please know that it is possible to feel remorse without stepping into hatred. Hatred, for yourself or anyone else, closes down the heart and stuffs one's grief, fear, and despair deeper into the body. You may argue that you're going to die anyway, but the question is how you want to die. Although we often have no control over how we are going to die, often we do.

Hatred consumes a great deal of energy, which could be used instead to intensify one's service role in the world, to focus on making the demise of other species more merciful, and to reap the benefits of the tools above that I have shared with the specific intention of supporting the reader in living and dying well in the last hours of our residence on the planet. Your hatred changes nothing, serves no one, and makes you abjectly miserable. It wastes your time and mine.

BECOMING A STUDENT OF EXTINCTION

In his beautiful 2014 article, "After the Harvest—Learning to Leave the Planet Gracefully," Robert Jensen writes the following:

> *The days of plenty are over, the high-energy phase of human life is coming to a close, and we have not yet learned all that we need to know—about ourselves or the world—to adapt to a new era. . . . As a people, we have yet to muster the intellectual resources, political will and moral courage needed to save ourselves and minimize the long-term damage to other living things. If that seems too much to bear, that's because it is. Yet that is our challenge: to face what is beyond our capacity to bear and refuse to turn away from the demands that these crises place on us.*

If the hospice patient is willing to gain the utmost from his/her time there, they will utilize it to reflect, reconcile, and ready themselves for death. Whether we choose to acknowledge it or not, we are meaning-making creatures. I don't know if elephants, dolphins, or chimpanzees, some of the creatures with more complex intelligences, make meaning, but humans either do or don't. Victor Frankl, one of the most famous survivors of a Nazi death camp, wrote in depth about his experiences in Auschwitz after being separated from his beloved wife who was murdered in another camp. In *Man's Search for Meaning*, which can be downloaded for free at numerous locations online, Frankl states that meaning came to him from three invaluable sources: purposeful work, love, and courage in the face of difficulty.

Extinction, like every other tragic event, offers us an opportunity to find meaning in it, and finding or making meaning is different for everyone. As we move closer toward the black maw of death, what sense do we make of that which seems so senseless? Making meaning does not necessarily feel "positive." In fact, it may feel inexplicably sad as we contemplate humanity's inability to fall in love with the Earth and partner with all aspects of the universe. However, our refusal to make meaning in the face of extinction is to rob ourselves of an essential aspect of our humanity. Humans have a capacity to experience a rich inner life, and when we shrug off the opportunity to find meaning as "insignificant," or when we argue that "there is no meaning," we silently consent to face our predicament in the same way that the majority of humans do: without consciousness, without compassion, and with the herd, as opposed to the heart, as the ultimate point of reference.

Overwhelmingly, when people contact me for life coaching, one of their implicit if not explicit questions is how one learns to live from their heart. Often they have read volumes of books and articles, watched numerous documentaries, and spent many hours a day online, yet they allow themselves to feel very little about nature and their relationship with it. When I ask them to take one hour per day to go outside and commune with nature and give them specific instructions for how to do so, they seem bewildered as to why I would make such a request. Yet almost without exception, when they actually do this, they report that they felt things in their bodies and psyches that they had never before experienced. Their hearts open, this has profound, usually salutary ramifications in their lives. In short, they are never the same.

Had most humans on this planet made the journey from head to heart, we would not be facing near-term human extinction. They didn't, and we are, and most humans will never make the journey. Many people tell me that there is nothing greater than the rational mind and the human ego. I then ask them how they deal with what I call "The Big Five": Love, death, suffering, the sacred, and eternity. They may argue that the last two don't exist. I may then ask if the first three have anything to do with the last two. Responses are mixed, but no matter what they are, I am quick to emphasize that we can't deal with the Big Five with the paradigm of Cartesian dualism and the rational ego so inordinately cherished by the fathers of the Enlightenment. In fact, the Enlightenment is strangely named because authentic "enlightenment" is trans-rational. Its paradigm simply cannot process the Big Five because it obviates and deplores mystery, and the Big Five is nothing if not the stuff of mystery. When mystery is eliminated, so are humility and the capacity to comprehend and live within human limits.

In these last hours, here are seven things we must be doing: 1) Falling in love with nature and allowing our hearts and bodies to weep with it; 2) Allowing that relationship to make exquisite meaning in our lives and determine everything we do; 3) Preparing with unprecedented awakeness, emotionally, spiritually, and logistically for death; 4) Doing everything humanly possible in our sphere of influence to practice good manners with other species and soften the impact of their demise; 5) Reflecting on our life in terms of how we lived it— what worked and what didn't work, what we wish had been "different, better, more"; 6) Making amends with people we have harmed; 7) Cultivating compassion for those who harmed us; 7) Creating beauty, magic, and joy in our lives as often as possible with as many people as possible.

Paradoxically, people report to me that when they practice these, they find themselves living more in their hearts and experiencing an unprecedented sense of vitality. As Michele Montaigne wrote, "To practice death is to practice freedom. A man who has learned how to die has unlearned how to be a slave."

CHAPTER 4
COMING TO GRIPS WITH DEATH

BY GUY R. MCPHERSON

All individual organisms die. When the last individual of a species dies, the species is extinct. All species go extinct. Although there is no reasonable counter to these obvious statements, most people refuse to believe humans will be extinct in the near future. In this death-denying, death-defying, omnicidal culture, the thought of our own death—much less our own extinction—is a bit too much for the typical citizen to bear.

Typical responses to the idea that humans will be extinct in the near future include the notion that it can't happen to us. We're too clever, too special. We'll find a way out. We always do. Another typical response: Why bother telling people? Why crush hope?

WHY BOTHER RELAYING EVIDENCE?

As Carl Sagan pointed out, "it is far better to grasp the universe as it really is than to persist in delusion, however satisfying and reassuring." Denying the evidence is persisting in delusion. We're not fans of that approach.

We see plenty of support for denying the obvious. Almost everybody reading these words has a vested interest in not wanting to think about climate change, which helps explain why the climate change deniers have won the battle for public opinion. According to a December 2013 paper in *Climatic Change*, the climate change counter-movement is funded to the tune of nearly a billion dollars each year.

And even as abnormal is the new normal, we're just getting started. Bruce Melton, writing for *Truthout* in a 26 December 2013 piece featuring climate scientist Wallace Broeker, points out, "Today we are operating on atmospheric concentrations of greenhouse gases from the 1970s. In the last 29 years we have emitted as many greenhouse gases as we emitted in the previous 236 years." In other words, the four-decade lag between emissions of greenhouse gases and consequences of those emissions is not complete. But it's on the way, and there is nothing to be done today to undo what we did during the last forty years. And, as pointed out by numerous scientific articles at a comprehensive summary dating back to February 2003 from the folks at Woods Hole Oceanographic Institution, abrupt and dramatic changes in climate aren't out of the question.

This knowledge brings with it horror and relief. I'm horrified by what's to come, which includes the near-certainty of human extinction by 2030 as we surpass 4 C above baseline. I'm relieved to know that today's consequences result from emissions dating to the mid-1970s, when I was excitedly learning to drive an automobile. I experience no teenaged guilt from youthful, ignorant actions.

I recognize that collapse of industrial civilization leads to a world 2 C warmer than baseline within a few days post-collapse, based on Clive Hamilton's assessment in his April 2013 book, *Earthmasters*. Where I live in the southwestern interior of a large continent in the Northern Hemisphere, that means we're headed for at least 5 C in the interior of large continents in the Northern Hemisphere shortly after collapse is complete. And that means no habitat for humans. The ensuing Dust Bowl will not end until long after humans have exited the planet.

Yet seemingly contrary to these simple, easy-to-reach conclusions, I work toward collapse. Largely unafflicted by the arrogance of humanism, I work on behalf of non-human species. Industrial civilization is destroying every aspect of the living planet, and I know virtually no one who wants to stop the runaway train. Yes, collapse will kill us. But our deaths are guaranteed regardless, unless I missed a memo.

I've given up on civilized humans making any effort to take relevant action. Never mind our stunning myopia. The money to be made is clearly more important than the extinctions we cause, including our own.

As pointed out in the March 2012 issue of *Nature Climate Change*, several psychological reasons explain why people have a hard time dealing with the stark reality of climate change (David Roberts comments at length in his 27 July 2012 article at *Grist*):

1) To the extent that climate change is an abstract concept, it is non intuitive and cognitively difficult to grasp.

2) Our moral judgment system is finely tuned to react to intentional transgressions—not unintentional ones.

3) Things that make us feel guilty provoke self-defensive mechanisms.

4) Uncertainty breeds wishful thinking, so the lack of definitive prognoses results in unreasonable optimism.

5) Our division into moral and political tribes generates ideological polarization; climate change becomes politicized.

6) Events do not seem urgent when they seem to be far away in time and space; out-group victims fall by the wayside.

At considerable risk of pummeling the dead equine, we'll reiterate several points mentioned in an earlier chapter (Chapter 2). Leading mainstream outlets routinely lie to the public. According to a report published 11 January 2014, "the BBC has spent tens of thousands of pounds over six years trying to keep secret an extraordinary 'eco' conference which has shaped its coverage of global warming." At the 2006 event, green activists and scientists—one of whom believes climate change is a bigger danger than global nuclear war—lectured twenty-eight of the BBC's senior executives.

Mainstream scientists minimize the message at every turn. As we've known for years, scientists almost invariably underplay climate impacts. I'm not implying conspiracy among scientists. Science selects for conservatism. Academia selects for extreme conservatism. These folks are loath to risk drawing undue attention to themselves by pointing out that there might be a threat to civilization. Never mind the near-term threat to our entire species (they couldn't care less about other species). If the truth is dire, they can find another, not-so-dire version. The concept is supported by an article in the February 2013 issue of *Global Environmental Change* pointing out that climate change scientists routinely underestimate impacts "by erring on the side of least drama."

In other words, science selects for conservatism (like picking cherries long after they are ripe). Science, after all, is merely the process of elucidating the obvious. Climate change scientists routinely underestimate impacts "by erring on the side of least drama" (like looking for the cherries long after they've fallen off the tree, onto the ground, and been consumed by rodents).

The feedbacks are too numerous, the inertia too strong. We fired the clathrate gun by 2007 or earlier, coincident with crossing the point of no return for climate change. The corporate media and corporate governments of the world keep lying, and too few hold them accountable.

Abrupt climate change is under way. Global climate change causes suffering and death of humans and other organisms. There is no escape.

WHAT ABOUT HOPE?

The hope card is trotted out as unimpeachable. It's nearly equivalent in its impact to playing the race card. Removing hope in this culture is analogous to denigrating motherhood and apple pie or taking candy from a baby (which doesn't seem like a bad idea from the perspective of personal health).

The definition of hope offered by Derrick Jensen is relevant: "hope is a longing for a future condition over which you have no agency; it means you are essentially powerless." In other words, hope is wishful thinking, also known as hopium. I'm certainly not willing to give up, and I constantly encourage acts of resistance that will allow opportunities for the living planet to persist into the future. In so doing, I'm channeling iconoclastic author Edward Abbey: "Action is the antidote to despair."

Hopium is the drug to which we're addicted. It's the desire to have our problems solved by others, instead of by ourselves. It's why we keep electing politicians while knowing they won't keep their promises, but finding ourselves too fearful to give up the much-promised future of never-ending growth on a finite planet.

Knowing we cannot occupy this finite world without adverse consequences for humans or other animals but afraid to face that truth, we turn away. We watch the television, go to the movies, gamble at casinos, play on Facebook, generally applauding while we take a flame-thrower to the planet. Nietzsche nailed it in his 1878 book, *All Too Human*, by saying, "In reality, hope is the worst of all evils, because it prolongs man's torments." Had he not been such a misogynist, Nietzsche might have included women, too.

Let's get off the crack pipe, and onto reality. May Pandora release the final gift from her container.

We are by no means suggesting the abandonment of (1) resistance or (2) joy-filled lives. Life, including human life, is a gift. Let's live as if we appreciate the gift. Let's live as if we appreciate the others in our lives, human and otherwise. Let's live as if there is more to life than the treadmill onto which we were born.

Let's live. In the spirit of seizing the day, let's live now. In the spirit of forward-thinking re-localization, let's live here now.

MOVING FORWARD

Capitulating to reality does not mean giving up. We've never suggested people give up on reality, resistance, lives of excellence, or love. We suggest pursuing each of these without delusion.

For us, resistance against the dominant paradigm is a moral imperative. Contrary to the popular expression, resistance is fertile, not futile. Whereas anybody can take the side of the heavy favorite—in this case, the side of the dominant paradigm—true courage and character are required to side with the underdog.

We're certain to die. Let's live.

We're certain to face extinction. Let's live with urgency.

Let's stop treating the entire planet as a series of resources to be exploited in the name of advancing civilization.

Currently, the typical perspective is to view finite substances and the living planet as materials to be exploited for our comfort. We treat resources as our entitlement.

Examples of intense anthropocentrism are so numerous in the English language it seems unfair to pick on this one word from among many. And, as with most other cases, we don't even think about these examples, much less question them: sustainability, civilization, and economic growth.

Let's start with definitions straight from the Merriam-Webster Dictionary:

> *Resource: 1 a: a source of supply or support: an available means—usually used in plural; b: a natural source of wealth or revenue—often used in plural; c: a natural feature or phenomenon that enhances the quality of human life; d: computable wealth—*

usually used in plural e: a source of information or expertise.

Each of these definitions implies an anthropogenic basis for resources, and c is particularly transparent on this point.

Digging a little further, the etymology of "resource" brings us directly to lifelong bedfellows anthropocentrism and Christianity. "Resource" is derived from the Old French *resourdre* (literally, to rise again), which has its roots in the Latin *resurgere* (to rise from the dead; also see resurrection).

From this etymology, it's a simple step back in time to Aristotle's final cause (which followed his material cause, efficient cause, and formal cause). Aristotle posited that, ultimately, events occurred to serve life, particularly the life of humans. This anthropocentric take on causality grew directly from the philosophy of Aristotle's teacher Plato, who focused his philosophy on separating humans from nature while popularizing the feel-good notion that humans have immortal souls. The idea that humans have souls, which was subsequently discredited by the (Western) science that grew from humble Grecian roots, became the basis for Christianity, one of three Abrahamic religions that developed in the Mediterranean a few centuries after Plato learned from Socrates and then taught Aristotle.

Considering the history of Western thought, it's no surprise we view every element on Earth as feedstock for industrialization. The only question for industrial humans is when we exploit Earth's bounty, not if. The logical progression, then, is to exploitation of humans to further feed the industrial machine.

Within the last few decades, personnel departments at major institutions became departments of human resources. Thus, whereas these departments formerly dealt with persons, they now deal with resources. There's a reason you feel like a cog in a grand imperial scheme. Not only are you viewed as a cog by the machine and by those who run the machine, but any non-cog-like behavior on your part leads to rejection of you and your actions. Seems you're either a tool of empire or you're a saboteur (i.e., terrorist).

It's clearly time to invest in wooden shoes.

As if even a small proportion of people in the industrialized world are willing to poke a stick in the eye of the corporations that run and ruin our lives.

Why is that? Probably because we think we depend upon them, when in fact they depend upon us. And, to a certain extent —to the extent we allow—we do depend upon industrial culture for our lives. But only in the short term, and only as self-absorbed, comfortable individuals unwilling to make changes in our lives (even ones that are necessary to our own survival).

Taking the longer, broader view, it is evident industrial culture is killing the living planet and our own species. The cultural problem we face is not that we're fish out of water. It's that we're fish in a river. We don't even know there's an ocean, much less a land base.

Aye, there's the rub. Evolution demands short-term thinking focused on individual survival. Most attempts to overcome our evolutionarily hardwired absorption with the self are selected against. The Overman is dead, killed by a high-fat diet and an unwillingness to exercise. Reflexively, we follow him to the grave.

Ultimately, we follow Nietzsche's Overman because we're tragically flawed organisms that, like other animals, lack free will.

Contrary to Descartes, Nietzsche concluded that our flaws define us, and therefore can't be overcome. We're human animals, hence far too human to overcome the tragedy built by evolution. Although we are thinking animals—what Nietzsche termed *res cogitans*—we are prey to muddled thoughts, that is, to ideas that lack clarity and distinctness. Nietzsche wasn't so pessimistic or naive to believe all our thoughts are muddled, of course. Ultimately, though, incompetence defines the human experience.

It's a short, easy step from Nietzsche's conclusion—we are flawed organisms—to industrial culture as a product of our incompetence. But the same step can be taken for every technology, with industrial culture as the potentially fatal blow. In other words, progress means only that we accelerate the rapidity with which bad things happen to societies, consistent with Jevons' paradox and its latest manifestation, the Khazzoom-Brookes postulate.

American exceptionalism thus becomes one more victim of the imperial train wreck that began when we first made tools.

PURSUING INTERPERSONAL GROWTH

Living with death in mind does not mean giving up. It doesn't mean becoming mired in despair. It doesn't mean capitulating to the shackles of culture.

Au, contraire. As pointed out by Goethe some two centuries ago, "None are more hopelessly enslaved than those who falsely believe they are free." Living with death in mind means liberation. It means breaking free from the shackles of culture. It means pursuing reality, resistance, lives of excellence, and love. None of these attributes are encouraged by the dominant paradigm, which instead promotes delusional thinking, wage-slavery, lives of inhuman ignorance, and misery.

Entering into a relationship with the natural world is one route to throwing off cultural shackles. When we move beyond viewing other people and other organisms as resources, we open ourselves to relationships that go deeper than exploitation. This, of course, will require love—love for nature, love for our children and grandchildren, love for each other. Love requires intimate knowledge. Let's get intimate with nature.

AND THEN WHAT?

As American ecologist Garrett Hardin pointed out long before his death a decade ago, that's the ecologically relevant

150

question. Anybody interested in individual or societal action must be willing to answer this question.

With respect to ongoing depletion of fossil fuels, any response to Hardin's question must include the matter of scale. Individuals are able to abandon fossil fuels before they abandon us. Doing so with grace is a bit challenging, but it's hardly impossible, as evidenced by numerous examples in the Transition Movement. Contemporary industrialized societies, on the other hand, are exhibiting little interest in adapting to a world without ready access to inexpensive fossil fuels. Apparently the people pulling the primary levers of industry would rather continue fighting than switch to a saner way of living.

With respect to ongoing, accelerating climate change, any response to Hardin's question must also include the matter of scale. Individuals are able to abandon a fossil fuel-fueled lifestyle with minor costs, including the disparagement that comes from living outside the mainstream. But, as illustrated by Jevons' paradox and the Khazzoom-Brookes postulate, individual choices do not translate to societal choices. An individual change in consciousness rarely leads to societal enlightenment. Jumping off the cruise ship of empire will not prevent the ship from striking the iceberg, and it nonetheless results in near-term death of the individual.

The following question then arises: What shall I do? How shall I live my life? In other words, now that we have knowledge of the near-term demise of our species, then what?

There are more than seven billion responses to the latter questions. Recognizing that birth is lethal and that we have an opportunity to demonstrate our humanity on the way out the door, we've chosen an eyes-wide-open, decidedly counter-cultural approach.

Beyond our own actions, we suggest individuals take actions they never previously imagined. We promote resistance against the dominant paradigm, even though—especially though—it is too late to save our species from near-term extinction. We propose assaulting ourselves and others with compassion. We recommend heavy doses of creativity and courage. We advise doing something well beyond the cultural current of the main stream.

At this point, what have you got to lose? Indeed, what have we got to lose?

CHAPTER 5

NEAR-TERM HUMAN EXTINCTION: THE MESSAGE IN THE MADNESS

BY CAROLYN BAKER

> *Before we can play with the angels of our compassion, we need to take tea with the 'demons' of our holding.*
>
> **~Stephen Levine, Healing Into Life and Death~**

This book is authored by two educators who no longer have formal positions as educators in institutions, but who, for a number of reasons, chose the profession. One of myriad reasons we chose to teach is that we have faith, no matter how severely it has been shaken by human-caused climate change, in the capacity of humans to learn. My enthusiasm about learning and teaching history was buoyed by a few occasions in the past where humans actually did learn something from history. Guy and I both miss teaching in the classroom, although we are still teaching in ways that we could not have predicted two decades ago. We enjoy stimulating and facilitating the learning process. However, the reader may wonder what purpose the process serves when history itself comes to a catastrophic end and when there is no more history or conservation biology to teach. Obviously, what it comes down to is questioning why we do what we do.

The physical evidence of catastrophic climate change is revealing itself so rapidly that a researcher like Guy is hard pressed to keep up with it. Even as this book goes to press, climate science will change with such dizzying speed that by the time you are reading these words, some of the research will be "old news." And what can we learn from it? Two things leap to mind. We can certainly learn what we as humans have done to bring about this cataclysmic nightmare. Additionally, we will surely learn that our lifespan has been greatly shortened by abrupt climate change and that we inhabit a planet in hospice. But that is hardly the end of the story.

Guy's research brings us the really bad news of our folly and our fate. The facts are brutal, and they hurt—terribly. However, it should not be assumed that my part of this book is going to soothe the pain or offer any sort of "comfort." My work has always been about making people as uncomfortable as possible. Why? Because humans do not learn anything from life and death situations unless they are compelled by circumstances to do so. In my years as a psychotherapist, never once did someone come into my office and say, "You know, I've made this rational decision that it would be a good idea for me to begin psychotherapy so that I could be a wiser, more loving, compassionate, whole person." Rather, almost everyone came screaming in emotional and spiritual pain. Humans were tragically, implacably unwilling to confront climate change until it was far too late. Even now, a ghastly number are still unwilling to even begin the process. This chapter is all about what we can learn as the Titanic sinks, but it will in no way eclipse the fact that the "unsinkable" ship is doomed.

So what is there to learn beyond how we screwed up and the fact that we're facing extinction? If we are pre-occupied solely with physical survival, there is absolutely nothing to learn. Let's eat, drink, and party like there's no tomorrow, let alone near-term human extinction. If your beating heart and your physical body are the only things that matter, then why are you reading these words? You should be out snorting cocaine or watching a "reality" TV show instead.

If, however, you sense that some part of you might be more significant than your physical body, your rational mind, and your human ego, you may want to keep reading. But I must give you fair warning as would any respectable television network showing a brutal massacre: Some of these words and verbal images may be disturbing.

Like the addict or alcoholic who has come to the very bottom of his addictive escapade—lying in a puddle of his own vomit in a rehab facility, hallucinating or imagining cockroaches crawling all over his body, wishing he were nothing but dead— any human being who is willing to face near-term human extinction is at the end of the line. There's nowhere else to go— not up, down, out, around, only in and through. This is what the bottomed-out addict finally realizes, but until she arrives there, intellectual masturbation rules, and abrupt climate change is just a bunch of charts and graphs that either "do not apply" to oneself or that really don't matter because "we're all going to die eventually, and that's the end of it."

The following are the real questions: What can you learn about who you are at your core, and what can you do to serve the Earth and make the extinction process easier for other members of it?

The following are the real questions: What can you learn about who you are at your core, and what can you do to serve the Earth and make the extinction process easier for other members of it?

TOP FIVE REGRETS OF THE DYING

Speaking for myself, I believe, and it has been my experience, that often people learn from their mistakes and even learn profound life lessons in the last hours of their lives. For example, Bronnie Ware is an Australian singer and songwriter who spent many years as a palliative care nurse. Her patients had gone home to die, but she was with them the last three to twelve weeks of their lives. Over the years, Ware noted the top regrets of the dying, which she compiled into a book aptly titled *The Top Five Regrets of the Dying*. The regrets are striking because they reveal the factors that, regardless of one's age or physical health, bring meaning and purpose to human lives, and those that do not.

An examination of each regret may be useful as we consider our place in history and the extinction in which we are now embroiled. Each regret has been seeded by the paradigm of empire and reveals the ultimate fruits that are harvested as a result of allowing the paradigm to grow in our lives.

1. I wish I'd had the courage to live a life true to myself, not the life others expected of me.

   According to Ware, this was the most common regret of all, and from my perspective is nothing less than archetypal, bringing to mind countless myths both ancient and modern from people who discover that the meaning of life eludes them. In the modern world in particular, we have few myths that actually provide a template for a life well lived. Throughout the twentieth century, and particularly at the conclusion of World War II, a popular myth took root in America that essentially stated if one worked hard, paid one's bills, and played by the rules, one would become a member of the middle class, and perhaps with exceptional effort become part of the ruling elite. We have come to call this the American Dream, however, the myth of the American Dream no longer serves and appears to unravel a bit more each day —a poignant symbol in current time, the economic demise of a city like Detroit, so woven into the fabric of post-World War II prosperity.

   Overwhelmingly, Ware found that dying people confessed that they had lived someone else's life and not their own. Whether it was the American Dream, the dutiful wife and mother, the loyal husband providing for his family, or the child graduating with honors to please her parents, Ware says that most of her patients did not live even half of their dreams. While it is far too late to re-start a career we wish we had pursued but didn't, we cannot begin to fathom the work that needs doing in terms of serving the Earth community on a planet facing the extinction of most life. I suggest making this the "dream" to which we aspire in the last hours of life on Earth

Overwhelmingly, Ware found that dying people confessed that they had lived someone else's life and not their own. Whether it was the American Dream, the dutiful wife and mother, the loyal husband providing for his family, or the child graduating with honors to please her parents, Ware says that most of her patients did not live even half of their dreams. While it is far too late to re-start a career we wish we had pursued but didn't, we cannot begin to fathom the work that needs doing in terms of serving the Earth community on a planet facing the extinction of most life. I suggest making this the "dream" to which we aspire in the last hours of life on Earth.

2. I wish I hadn't worked so hard.

For me, the most striking aspect of this statement is that no one said they wished they had worked harder! Most men, of course, wished they had spent more time with their families, although some women regretted working too much as well. Traditionally men, more than women, find their identity in work. Men are more likely to be socialized to be "human doings" rather than "human beings," and while women are traditionally expected to tend to the family, men are expected to work outside of the home and focus on providing rather than nurturing.

Our assumptions about work have been formed by the paradigm of the old money system that is based on separation and competition. In *Sacred Economics: Money, Gift, and Society in the Age of Transition*, Charles Eisenstein has given us a template for the Gift Economy and how to implement it. In that paradigm, we may choose to work hard, but not for the same reasons we now do, which have to do with competitive, ruthless survival and the terror of scarcity. Moreover, in the new paradigm of the Gift Economy that Eisenstein says is "about creating the more beautiful world that our hearts know is possible," rest, relaxation, celebration, joy, and play must be woven into our work. The more they are, the more the lines between work and play are blissfully blurred as we sense with our bodies that labor and laughter are not enemies but long-lost friends who cry out to be integrated in the psyche of the human going extinct.

3. I wish I'd had the courage to express my feelings.

According to Ware, many people suppressed their feelings to keep peace or to please others, or because they minimized the importance of their emotions. "As a result," says Ware, "they settled for a mediocre existence and never became who they were truly capable of becoming." Additionally, many developed illnesses relating to their unexpressed bitterness and resentment.

Civilization convinces most of its inhabitants that emotion is essentially irrelevant and unequivocally secondary to thinking. What matters, it tells us, is not what you feel but what and how you think. While this notion is prevalent among men, women buy into it as well. In empire, we tragically live most of our lives minimizing emotion. Then, at the end of our days, we regret that we didn't value it until then. Near-term human extinction is guaranteed to "remedy" our resistance to emotion. If you are reading these words, you have probably already experienced many emotions related to extinction and will be experiencing many more in the days ahead.

4. I wish I had stayed in touch with my friends.

According to Ware, many people suppressed their feelings to keep peace or to please others, or because they minimized the importance of their emotions. "As a result," says Ware, "they settled for a mediocre existence and never became who they were truly capable of becoming." Additionally, many developed illnesses relating to their unexpressed bitterness and resentment.

Civilization convinces most of its inhabitants that emotion is essentially irrelevant and unequivocally secondary to thinking. What matters, it tells us, is not what you feel but what and how you think. While this notion is prevalent among men, women buy into it as well. In empire, we tragically live most of our lives minimizing emotion. Then, at the end of our days, we regret that we didn't value it until then. Near-term human extinction is guaranteed to "remedy" our resistance to emotion. If you are reading these words, you have probably already experienced many emotions related to extinction and will be experiencing many more in the days ahead.

5.   I wish that I had let myself be happier.

Throughout my writing in recent years, I have frequently made the distinction between happiness and joy. Essentially, happiness is a transient state that depends on external circumstances. If most are positive, we tend to feel happy; if most are not going well, we tend to feel unhappy. Joy, however, is a state of deep contentment that can be present even when we are under extreme stress, experience a devastating loss, or undergo a major life transition. Happiness is usually connected with how the ego perceives its situation whereas joy has more to do with one's connection with the eternal or the sacred self within.

My sense is that when dying people state that they wish they had allowed themselves to be happier, they are really saying that they relied mostly on the barometer of external circumstances to dictate their mood but did not experience the deep contentment inherent in grounding one's consciousness in the deeper self. To ground in the deeper self and experience a profound sense of joy, one must attend to the four issues above that can produce regret at the end of our lives.

Bronnie Ware notes that those who did not allow themselves to be happy often "pretended to others and to themselves that they were content, when deep within, they longed to laugh properly and have silliness in their life again."

What do you regret so far? What are you willing to do to minimize your regrets? If we continue to live out the paradigm of industrial civilization, we are likely to accumulate many more regrets. If we are willing, however, to forge and embrace a pre-extinction paradigm, we may not only experience fewer regrets, but also harvest the fruits of a meaningful life well lived.

THE PRE-EXTINCTION PARADIGM: EXORCISING EMPIRE

If you've read Guy's book *Walking Away from Empire*, you will learn of his personal journey of leaving a tenured professorship to radically alter his living arrangements in preparation for the collapse of industrial civilization. I've thoroughly enjoyed this touching, inspiring, thought-provoking, sometimes snarky, sometimes heartbreaking saga of awakening and courageous abandonment of civilization's paradigm. (We include this topic in our "Conversations" section below.)

Yet throughout my reading of the book, one question would not relent: Is it really possible to walk away from empire? In my dialogs with Guy, I discovered that he would be the first to agree that for a variety of reasons walking away from empire is not possible. In dialog with myself, I realized that the tentacles of empire reach so far into my own psyche and have entangled themselves so deeply that I am profoundly limited in the extent to which I can walk away, yet at the same time, I believe that we all must make every attempt to do just that.

For me there are two overshadowing obstacles to exiting empire, both of which are related to the internal dynamics of empire programming. They are so profound that, on one level, radically altering one's living arrangements may be the least daunting facet of making the break.

ENLIGHTENMENT ENCULTURATION

The first of these two obstacles is Enlightenment Enculturation. The Enlightenment, that intellectual about-face that occurred in the seventeenth and eighteenth centuries in the West following what we now call the Dark Ages, was committed to eradicating the ignorance and superstition perpetuated by the Roman Catholic Church and folk wisdom. On the one hand, the Enlightenment was a breath of fresh air when compared with commonplace beliefs that women and black cats caused the Black Death of the fourteenth century and the church's implacable insistence that the Earth, not the sun, was the center of the universe. On the other hand, and equally implacably, the Enlightenment committed itself to one path of knowledge only, namely reason. In doing so, the Enlightenment, in part, set in motion the paradigm of industrial civilization which glorified logic and the masculine, disparaged intuition and the feminine, and instituted a way of living based on power, control, separation, and resource exploitation. Ultimately, how different the rule of this paradigm was and is from the hierarchical, fundamentalist domination of the church is arguable.

One of the few places in Guy's extraordinary book with which I must take issue with is this same dichotomy—which I believe to be a false one—that is, a dichotomy between reason and mysticism. Curiously, the intellectual giants of Classical Greece whom most modern thinkers admire, were deeply mystical. The word mysticism is related to mystery, and very specifically, to myth or mythology in which Classical Greek thinkers had been steeped since birth. Myths were sacred narratives for the Greeks that served as models for behavior. The predominant theme of all myths of their time was the notion that humans were not superior to the gods and goddesses and that as soon as they attempted to be, they would experience some aspect of personal or communal demise. For me, mysticism does not eclipse scientific inquiry, but instead may complement it in many instances. (Please see the conversations section for our dialog on this topic.)

Author Peter Kingsley has written extensively in his four books, *Reality: A Story Waiting to Pierce You, In the Dark Places of Wisdom*, and *Ancient Philosophy: Mystery and Magic*, of the likelihood of widespread contact between Ancient Greek philosophers and sages of Eastern philosophy. In an article titled "The Paths of the Ancient Sages: A Sacred Tradition Between East and West," Kingsley documents instances of contact which are overwhelmingly excluded from traditional histories of philosophy in the West. The Western philosophical tradition has attempted to surgically remove accounts of the interpenetration of East and West in the Ancient and Classical Greek eras, but more extensive research reveals that for philosophers such as Pythagoras, Parmenides, and Empedocles to name only three, knowledge was as much about direct intuitive, physiological experience as about intellectual understanding.

Hundreds of years later in the twentieth century, psychologist Carl Jung began writing about the four functions of consciousness: thinking, feeling, sensation, and intuition. Jung theorized that although everyone has a dominant function, as well as an inferior one, if we exclude any function or fail to develop it, imbalance results and we become one-sided individuals. At approximately the same time, Katherine Cook Briggs and her daughter Isabel Briggs Myers devised a fairly reliable personality type indicator. The Myers-Briggs inventory is a useful assessment of personality and how we construe our experiences. All personality types have strengths and weaknesses, and knowledge of the types can prove extremely useful in both personal and community relationships.

For me, Jung was the ultimate reasoned mystic, as were his contemporaries such as Albert Einstein, David Bohm, Werner Heisenberg, and Erwin Schrödinger. If any humans are around a few decades from now, they will be incapable of forging a human existence that radically departs from our own without the integration of the rational and the sacred.

Enlightenment Enculturation can be particularly damaging if we exclude other functions besides thinking from our interpersonal relationships. For example, if one is a thinking type, relying primarily on reason and intellect, one will need to work harder at intuiting a situation, identifying and expressing one's feelings about it, and noticing the sensations that occur in the body during interactions with others. The classic situations where I have witnessed this challenge are among members of a living community, a regional community, or with people in romantic partnerships. I repeatedly encounter individuals who are working together on collapse preparation or community building projects and are endeavoring to proceed primarily from a thinking-type perspective, as if reason and logic alone can solve problems and resolve all adversity.

For example, let's say that a man named Joe works very hard at being reasonable and analyzing situations logically, but he may not have noticed or even heard the tone of voice with which Nancy in the group has uttered a response to Joe's comment. Frank, who's very intuitive, has sensed a potential conflict brewing in the group, and Frank's wife, Vivian, a sensate type, may have experienced a sharp sensation in the pit of her stomach during the conversation and perhaps later, a sense that something was "off." None of these individuals need verbalize their responses in the moment, but they absolutely must pay attention to them. Hopefully, they have learned or are learning some solid dialoging skills. Otherwise, their collaboration will probably be short-lived.

I never tire of emphasizing the need for emotional literacy and communication skill development in preparing for and navigating the collapse of industrial civilization or near-term extinction. The more I work with groups and individuals who are preparing, the more I witness how woefully unprepared most of us are for dealing with the non-logistical aspects of attempting to walk away from empire.

Passionately echoing my missives, is Charles Eisenstein's assessment of the limits of reason from his *The Ascent of Humanity*:

Reason cannot evaluate truth. Reason cannot apprehend beauty. Reason knows nothing of love. Living from the head brings us to the same place, whether as individuals or as a society. It brings us to a multiplicity of crises. The head tries to manage them through more of the same methods of control, and the crises eventually intensify. Eventually, they become unmanageable and the illusion of control becomes transparent; the head surrenders and the heart can take over once again.

The positive legacies of the Enlightenment are many: Learning to think rigorously and critically, questioning authority, freedom from the impediments of superstition, and reveling in the delights of understanding our world and making sense of it. Yet Enlightenment Enculturation has become yet another face of fundamentalism in the last four hundred years as a result of its intractable insistence that reason is the only valid method for coping with the vicissitudes of the human condition. For me, Jung was brilliant not only in his assessment of the four functions of consciousness but also in his realization of the value of the dark and irrational aspects of humanity.

EMPIRE'S STICKY SHADOW

The other overshadowing influence that prevents us from extricating ourselves from empire is its shadow. Of the myriad contributions of Jung, one of the most significant was the concept of the shadow. While indigenous people had been well aware of the notion for millennia, few Westerners were at the time Jung began writing about it in the twentieth century. Overall, the shadow means everything that lies outside of consciousness, which may be positive or negative. The shadow is usually the polar opposite of what we perceive as true about ourselves. For example, part of us is committed to leaving empire and radically changing our living arrangements, but another part resists doing so. Or on the one hand we despise the entitlement we see around us in our culture, yet some part of us feels entitled, and if this part of us is not made conscious, it can sabotage our efforts to leave empire or manifest as entitlement within the parameters of our new living arrangements. Any aspect of the shadow can in fact surface unexpectedly and unconsciously sabotage us or harm another individual or group that we consciously cherish.

We may proclaim our desire to join with others in a living community or a group endeavor, but some part of us actually resists joining and will find a way to undermine a person or a project. This may manifest in myriad ways, including hyper-criticism, passive-aggressive behavior, blaming, adopting a victim stance, or even abandoning the group.

Changing our living arrangement is but one small first step in the journey away from empire. The "well adjusted" citizen of empire abides with us wherever we go or alongside everything we do in order to live the new paradigm. Constant introspection, not of the obsessive variety, but deep reflection and conscious intention to make conscious our residual shadow is imperative for ex-patriots of empire. Our new living arrangements will more than likely catapult the shadow to the surface, and it will be much better for us and everyone else if we know that and work with it in advance.

Journaling is an excellent tool as well as working with polarities. In my forthcoming book, *Love in the Age of Ecological Apocalypse: The Relationships We Need to Thrive*, I provide specific journaling tools for working with shadow polarities that can be profoundly helpful in bringing the shadow to consciousness and clearing it.

In *Dispelling Wetiko*, author Paul Levy dives deep into the wounds of American culture and assists the reader in recognizing and healing the wounds within oneself and in the community. "Wetiko," a Native American word, simply means "a diabolically wicked person or spirit who terrorizes others by means of evil acts."(3)   Levy names the collective psychosis "Malignant Egophrenia" (ME disease), which he defines as a condition in which the human ego is estranged from the deeper Self. The Self, according to Jung, is the sacred within or that part of the psyche that is eternal and connected with something greater than the rational mind and ego. A principal aspect of Jungian psychology was and is the establishment of a relationship between the ego and the Self in order to transform consciousness and enhance wholeness. This requires a journey into the unconscious and most importantly, encountering the shadow and its projections.

Levy describes what happens when we do not develop a relationship with the shadow. (p.99) When we do not develop a relationship with the shadow, we end up projecting it outward—a phenomenon that is at the root of the ME or wetiko disease. While it may be easy to identify the shadow projections of the rich and powerful and how they blame "the evil-doers" or a particular ethnicity for the world's ills, it is more difficult to recognize shadow projections within ourselves. To help illumine us, Levy offers assistance in how we can recognize and befriend the shadow and at the same time "become more immune to moral decay and psychic infections such as wetiko." He emphasizes that "if we learn to deal with our own shadow, we have truly done something real and of significant benefit for the whole world."

The disease of "malignant egophrenia" (ME) is also the disease of industrial civilization, an economic, political, and social arrangement that requires violence to maintain itself. Every inhabitant of industrial civilization is infected with the ME disease, but Levy notes (p.13) that "full blown" Wetikos "are not in touch with their own humanity and therefore can't see the humanity in others."

The acronym "ME" is certainly no accident because malignant egophrenia naturally causes psychic vision to focus on "me" and my needs rather than more broadly on the rest of the world in addition to me. When we are unwittingly victims and carriers of the ME disease, we perpetuate the collective psychosis and the phenomenal evil of which it is capable. The most authentic and salutary response to near-term extinction is a commitment to heal our ME disease by falling in love with the Earth and serving all members of its community.

With all due respect to Charles Eisenstein's assertion that evil does not exist and that we assume it does because the belief is inculcated in us by empire, I must disagree. In today's milieu, many who bristle at what they assume is its religious undertone avoid the use of the word "evil." They prefer instead to use words like "misguided," "fallacious," "perverse," or "odious," but certainly not "evil." Yet do those words accurately describe the atrocities against the Earth and against our own species? Take for instance such examples as the Sandy Hook massacre, the carnage of terrorism, the bloodbaths of the Iraq and Afghanistan wars, drone attacks on innocent civilians, human trafficking, the institutionalized sexual abuse of children by Roman Catholic and other clergy, the rampant assaults on the Earth and humans by hydraulic fracturing, the covering up of the severity of the Fukushima nuclear power plant accident, the massive, rampant, institutionalized corruption of Wall Street and corporate capitalism worldwide. Evil is the only appropriate term for these abhorrent acts. The following is Levy's account:

> *Our lack of imagination for the evil existing in potential in humanity is a direct reflection of a lack of intimacy with our own potential for evil, which further serves to enable the malevolence of wetiko to have nearly free rein in our world. We can't afford to have a concept of evil that is too small. (14)*

In other words, part of the reason we are where we are in human history is that we have not imagined the scope of evil that humans can fall prey to. Moreover, contrary to Eisenstein's assertion that the origins of the notion of "evil" lie within empire itself, countless indigenous cultures over the course of millennia have in fact understood the reality and the pitfalls of evil. Many have developed elaborate rituals for protecting their communities from evil and have adopted lifestyles that conscientiously avoid harm to any members of the Earth community.

And while it is important to name evil in the world, its eradication begins with ourselves. We need to become intimately acquainted with our own shadow and the difference between what Jung called the daemon in us and the demon.

"The daemonic," says Levy, "is the urge in every human being to affirm itself, assert itself, and perpetuate itself; it is the voice of the generative process within an individual." (16) Our "daemon" refers to our gifts and innate talents. Moreover, the channels available to us in industrial civilization for utilizing and fulfilling the daemonic are denied at worst and stifled at best. However, if the daemonic is not honored and treated as sacred, it becomes demonic and destructive. Thus, loving our creativity and nurturing it is an enormous asset in transforming both internal and external darkness.

Our greatest protection, according to Levy, is when we are in touch with our true nature, our inherent wholeness that Jung called the Self. The real "cure" for wetiko from Levy's perspective is a radical shift in consciousness and awareness that "there is no place to take refuge, except in the true nature of our being." (40) The result is a new kind of logic that knows that interdependence, unlimited wholeness, and the unity of all things constitutes the framework of a new paradigm that liberates us from the old story of industrial civilization and signals the termination of our engagement with the collective psychosis.

From the Jungian perspective, humans possess a personal unconscious but are also part of a collective unconscious. Levy recounts the following (47):

> *Whenever the contents of the collective unconscious become activated, they have an unsettling effect on the conscious mind of everyone. When this psychic dynamic is not consciously metabolized, not just within an individual but collectively, the mental state of the people as a whole might well be compared to a psychosis. Jung never tired of warning that the greatest danger that threatens humanity is the possibility that millions of us can fall into our unconscious together and reinforce each other's blind spots, feeding a contagious collective psychosis in which we unwittingly become complicit in supporting the insanity of endless wars; this is unfortunately an exact description of what is currently happening.*

Another way of viewing the dispelling of wetiko is that each of us must "un-crazy" ourselves in a world in which we are engulfed in a psychic epidemic that appears totally normal to its victims. In my article "Maintaining Mental Health in the Age of Madness," I argued that the mental health community is also a casualty of the collective psychosis and remains so embedded in it that it has become yet another "carrier" of the disease, with little to offer those infected by it unless mental health caregivers are themselves attending to the personal and collective shadow.

While we can engage in endless analyses of the external darkness, how do we bring the light of consciousness to the darkness within each of us? "Self-reflection," says Levy, "is not only the most beneficial response to evil, it is in fact the only response where we have any real influence or control." (159)

As we wake up to how we are participating in the insanity, we are more available to connect with others, even if they do not completely agree with us. In a sense, we become as Levy says, "islands of sanity in an ocean of madness. Over time the islands in our archipelago of sanity join with each other and form continents, so to speak, as we dispel the madness in the field." (178)

RESPONDING TO THE MESSAGE IN THE MADNESS

Notwithstanding the top five regrets of the dying, most of us have many more than five. We must be willing to unflinchingly bear witness to the parts of ourselves that have contributed to the omnicide that has consigned us to hospice. Many books and chapters outlining our planetary predicament end with a "What You Can Do" section that supposedly compensates the reader for the unsettling emotions the information may have produced. I end this chapter with a "What You Can Do" section, but it is not meant to "console" the reader. It is instead intended to unsettle you even more, yet my wish is that you will choose to take action anyway.

1. Be willing to recognize that you are now a patient in hospice. You have only a certain amount of time to continue living on this planet. You have no control over when you will leave it. You cannot reverse catastrophic climate change. Your children and grandchildren are not likely to outlive you.

2. Begin living your life as a hospice patient might because no other way of living any longer makes sense. Every relationship, every act, every experience is deeply meaningful. Live it as if it were the last. Live with passion and be as fully present as possible in every situation. Savor your aliveness and vitality.

3. Commit to a time of self-reflection daily. Consider journaling about your life—your relationships, your past contributions to the Earth community, what worked for you and brought you joy and meaning, what did not work and brought you sorrow and regret. Practice a meditation technique daily.

4. Make a list of people to whom you would like to make amends. Speak to them directly if possible. If not possible, write them a letter that you cannot or choose

not to send because making the amends is really for you as much as it is for them. (You don't need to tell them that you are in hospice.)

5. Make a list of people whom you want to thank for what they have contributed to your life. (You don't need to tell them that you are in hospice.)

6. Deeply evaluate your life in terms of your time commitments and activities and structure at least several hours a week of service into your routine. If you are in a service profession, notice what else you might do beyond what you are being paid for. Make extinction as easy on other species and other humans as possible.

7. Spend at least one hour per week in nature as suggested above. Allow yourself to wander with your senses fully engaged. That is, watch, listen, smell, taste, and touch the many facets of nature you encounter. Wander with your senses, not your mind. Experiment with putting aside your rational mind long enough to engage in a conversation with nature from the heart. Ask a question and then listen. Speak gratitude to the beings you encounter in nature and spend some time apologizing to nature for how our species has harmed it. Allow yourself to grieve deeply, perhaps repeatedly if necessary, for what humans have done to the Earth, including your participation in the omnicide. Consider that the heart of the Earth may be broken and that it wants and needs your grief.

8. Immerse yourself in love—both giving it and receiving it. Be aware of what keeps your heart closed to certain people and situations. In 2012, Guy McPherson wrote a powerful article titled "Only Love Remains." If you are walking around with contempt for the human species, you are unable to give and receive love. Deal with the grief that lies beneath your rage. The steps above will assist you in doing so.

9. Immerse yourself in fun, play, and joy. Yes, you're in hospice, and yes, you're going to die, but people in hospice can actually have fun. Allow yourself to laugh, and "inflict" joy on others.

10. Create beauty. You are wired for enjoying art, beautiful music, gorgeous sunsets, and the heart-melting colors of flowers. Share beauty with others. You don't have to be an artist or a musician to create beauty. The poet Rumi wrote, "Let the beauty of what you love be what you do."

Only time will reveal how precious these present moments are. And as another mystical Middle Eastern poet Hafiz reminds us, "Now is the time to know that all you do is sacred."

CHAPTER 6
CONVERSATIONS WITH GUY AND CAROLYN: LIVING WITH NEAR-TERM HUMAN EXTINCTION

HUMOR

Carolyn: Guy, I recall visiting you last summer at the Mud Hut, and one of the things that impressed me was your sense of humor. I also observed this when you came to Boulder a few months later. I suppose you've always had a sense of humor, but I'm wondering how you keep it when you pretty much dig up the latest dire climate research every day. I'm also wondering how it serves you.

Guy: It's difficult to pinpoint the birth of my sense of humor, although I know it came early. My dad instilled in me a talent for telling stories, and I grew up a bookworm, therefore a misfit in a redneck, backwoods logging town. Possessing a sense of humor allowed me to avoid a few beatings at the hands of bullies.

I had developed a wry, self-deprecating sense of humor by the time I finished high school. It allowed me to fit into a culture with which I was uncomfortable without attracting undue attention to myself. With respect to the latter phenomenon, I didn't raise my hand in a classroom between high school and the final year of my Ph.D. program specifically because I dreaded the attention of others.

I raced through undergraduate and graduate degrees in resource extraction—an undergraduate degree in forestry was succeeded by M.S. and Ph.D. degrees in the mundane field of range science. Thus, I learned how to grow large trees quickly, and then moved onto exploitation of other terrestrial ecosystems. A quick stint as a postdoctoral researcher at the University of Georgia was followed by a teaching position in the Range Science Department at Texas A&M University at the tender age of twenty-eight. I was offered a tenure-track position at the University of Arizona even before I commenced the position in College Station, Texas, but I spent a college year at Texas A&M before embarking on my tenured path in the Sonoran Desert.

The position at Texas A&M was critical to the development of my career and also to the development of my sense of humor. I was hired primarily to teach an ecology course for non-majors. It was targeted at second-year students. The course acted as a "core" science course for undergraduate students, hence allowing the clever sophomore to escape courses such as chemistry and physics. Not only did I face a lecture hall filled with indifferent students three days weekly, I was allowed to repeat the experience each teaching day while also organizing eighteen laboratory sections, hiring and supervising graduate students to teach seventeen of the sections, and teaching one of the sections myself.

A typical week therefore began early Monday morning in a lecture hall with me and 150 half-awake nineteen-year-olds who desired only a passing grade on their way to fame and fortune in the worlds of business, fashion, and petroleum extraction. Campus culture included bringing the daily campus newsletter to the morning course and, if the class wasn't thoroughly captivating within the first few minutes, opening the newspaper. Few sights are more demoralizing than lecturing to a sea of crackling newspapers.

Thus was born my career in stand-up tragedy.

I would regale poor, unsuspecting business majors with tales of woe induced by their chosen field. And I would make them laugh. I excelled at edu-tainment.

Afternoon brought another section of the same class, this one numbering 250 students. Similarly disinterested in the topic, this group lacked the crackling journalistic weapons characteristic of the morning crew, but it suffered post-lunch fatigue and the expected skepticism with respect to activities that might interfere with careers focused on the acquisition of fiat currency. "The business of America," as US President Calvin Coolidge had pointed out six decades earlier, "is business."

Between the two large lecture events was a laboratory section of the same course. I hauled young women in short skirts into the tangled, filthy web of reality to discover the glories of the natural world. A large dose of humor was required to see all twenty of us through the three-hour ordeal.

Tuesdays and Thursdays brought introductory courses in rangeland management. Every otherwise spare moment during the week was filled with visits to the laboratory sections taught by the ten graduate students, as I was expected to turn each of them into a stellar teacher. Between their classes and mine, I met with them one-on-one to nurse their tender egos and explain that teaching couldn't be mastered in a single semester. I adamantly expected them to learn the names of every student in their section with the first few weeks, and I held myself to that standard for all 400 students.

With the exception of a single course I taught to a few peers late in my graduate career, this was my first semester in the classroom. The cliché "trial by fire" comes to mind.

By the time I arrived at the University of Arizona in May of 1989, I was a seasoned teacher and mentor. During the following two decades, I supervised dozens of graduate students and taught hundreds of students. I could barely stand teaching the same material twice, so I didn't. I updated my courses upon every offering, and familiarized myself with each student before the semester was two weeks old. After all, I knew students don't care what the teacher knows unless they know the teacher cares. So I took time and effort to care about my students, and infused a healthy dose of self-deprecating humor into every session in the classroom.

That's a lot of background material to answer the seemingly simple questions you asked.

With respect to the dire information about climate change, I've come to view my own life as absurd. Ditto for industrial civilization. The words of Camus come to mind: "The only way to deal with an unfree world is to become so absolutely free that your very existence is an act of rebellion." My goal, at this late juncture in my life, is to follow Camus' lead in treating my every act as one of rebellion.

I hate civilization, yet I'm entrenched with it. I particularly hate industrial civilization, yet shortly after it collapses there will be no habitat where I live. I hate patriarchy, and I'm a white male at the apex of patriarchy. The odds against any one of us being here on this most wondrous of planets are astronomical. And yet here we are. Here I am. The absurdity is profound.

With respect to the issue of how my sense of humor serves me, I have no choice. Like other humans, and other animals, I have no free will. So I'm left with my sense of humor, which often reveals itself at socially inappropriate times.

Given a choice, I wouldn't give up my sense of humor. I've staked out a lonely position within the scientific community. I can't imagine I would survive without viewing the universe, and my place within it, as absurd. I can't imagine I would be able to relay my message about near-term human extinction, and survive the ordeal, in the absence of a good sense of humor.

What about you? How do you come to grips with the dire news you're transmitting to others?

Carolyn: I grew up in a bible-thumping, fundamentalist Christian home in Indiana which most would assume would be devoid of any humor at all. Nevertheless, amid the Calivinism that could not be cut with a chainsaw, my dad did have a sense of humor. Occasionally, it would escape his vigilant control, and laughter was permitted, especially if it was initiated by his jokes.

At Michigan State University, I escaped all things religious and discovered the joys of sixties activism and recreational drug use. Not surprisingly, the first time I smoked marijuana I felt as sober as Calvin himself, adamantly denying that I felt differently than before I started smoking. Unlike Bill Clinton, I inhaled. Nothing's happening, I tell you. Nothing's happening. Meanwhile, a cluster of friends surrounding me were falling off their chairs laughing as I blatantly demonstrated how not-unaffected by the magical weed I was.

Throughout my adult life my sense of humor honed itself and became increasingly snarky, particularly alongside any confrontation with the industrial civilization that you and I both hate.

Like you, I taught college students and was an adjunct professor of history and psychology. I had the most fun teaching history because at the time, the US occupation of Iraq was in full swing, and I was living near one of the largest Army installations for deployment to Iraq, namely Fort Bliss in El Paso, Texas. I taught at a community college and two nearby universities, but the real gem was teaching US history for the community college at Fort Bliss itself. The first hot summer night I drove up to the main gate and encountered the MP's, I was shaking in my sandals. Upon entering the classroom I was deluged with students who insisted on calling me "ma'am," and I was certain that the kind of US history I would be teaching would get me fired at best. The words "indefinite detention" kept reverberating in my mind.

I began teaching these students my "Howard Zinn on steroids" version of US history, using lots of snark, and to my surprise, my students loved it. Gee, why was I surprised? If you're black, Hispanic, or a working class white man or woman getting ready to deploy to the Middle Eastern death machine, why would you not feel relieved when someone begins to spell out for you how this happened?

I loved my military students, and they loved me. In fact, one day while walking down the hall in the social science building, I heard the gruff Brooklyn voice of my faculty supervisor calling out, "Hey, Carolyn." I'm thinking, "Oh, crap, what have I done now?" Tentatively, I turned around to respond, thinking that I was certainly in trouble. He told me that the Fort Bliss students loved me and that he wanted to give me more classes there. I didn't know whether to smile or cry, but I kept teaching at Fort Bliss and other campuses where, for the most part, students couldn't have cared less about history or else found my comments about the Iraq War "offensive." I was reported to administration once for "traumatizing" an Iraq vet in one of my classes. Turns out he wasn't traumatized at all, but my supervisor on that campus got wind of my highly vocal opinions and told me to be aware of how conservative the community was. Within a year after that incident, relatives of my students began returning from Iraq in body bags, and my version of US history became increasingly popular among the students. It reminds me of what you and I are speaking and writing about abrupt climate change. It's all very "extreme" until a hurricane hits your house or a drought begins killing your livestock. During my teaching career, some of my colleagues were excellent allies, and we supported each other often with shameless humor—of both the light and dark variety.

Meanwhile, many of my students were getting their news from John Stewart or Stephen Colbert. While I never spent much time watching either show, I feel indebted to Stewart and Colbert for making my job a bit easier by cultivating in my students a level of openness to the snark I was dishing out in my classes.

Today, I temper tragedy with a number of things, humor being one. I have published a subscription-based *Daily News Digest* for seven years that gives me a platform for pontificating and also injecting a daily dose of humor into the news I'm posting. Friends constantly send me jokes and links to humor online. I particularly love the *New Yorker*'s Borowitz Report.

Along with humor, I find that I require a certain amount of play each week. Whether it involves playing with my dog or friends or escaping into a TV series like "Breaking Bad," I can't spend sixteen hours a day with my face in a computer. And no, I don't have a garden or a Mud Hut to tend, but when I want to really get away from the computer, I go to my favorite local park, making sure to leave my cell phone at home, and just allow myself to "veg out" with nature. The joy I find there is incredibly restorative.

Carolyn: And so, Guy, I'm thinking that our readers might like to hear something from you about how you restore and recharge in the midst of doing the very challenging work you do. I know you love it, but I also know what a burden it can be to know what we know.

Guy: From American spiritual teacher Adyashanti comes this wisdom: "Make no mistake about it—enlightenment is a destructive process. It has nothing to do with becoming better or being happier. Enlightenment is the crumbling away of untruth. It's seeing through the façade of pretense. It's the complete eradication of everything we imagined to be true."

At least for me, awareness is not a gift. It's a curse. I have become a pariah with increasing radicalism (i.e., getting to the root).

My antidote, paradoxically, is increasing radicalism. In for a penny, in for a pound, as the expression goes. I'm fully committed to rooting out the evidence. I'm thrilled when I'm able to find a little-noticed bit of evidence, and even more thrilled when I'm able to weave the bit into a compelling, comprehensive story. It's a salve on my psyche.

I love the natural world, too. My mind can hold two thoughts simultaneously: (1) We have destroyed, and continue destroy, habitat for humans and other organisms at an astonishing rate, and (2) life is wonderful. My love for the natural world, and my frequent immersion into it, serve to restore my spirit. To observe the stunning resilience of the natural world in the wake of industrial civilization is truly awesome.

Finally, my travels bring me into contact with like-minded people. These people understand the dire nature of our collective straits. They understand that industrial civilization is responsible for fouling the air, dirtying the water, and driving the Sixth Great Extinction. They understand that we're taking ourselves into the ultimate abyss. They understand, to a far greater extent than my former colleagues, the cliff to which we're accelerating. And they're filled with joy and good humor.

These good people don't pursue happiness; it finds them because they are not seeking it. These are caring, committed individuals who aim to be fully present, hence fully engaged in relationships that perhaps began moments ago. They stimulate my senses and my heart in a manner similar to the joy I experience when I unravel a scientific mystery or immerse myself in nature's bounty.

For me, science, nature, and people represent the trifecta of the human experience. They bring balance. They remind me I'm fully alive. Indeed, they make me fully alive.

Carolyn: I resonate deeply with the words of Adyashanti. At an early age, my bullshit detector began honing to near perfection. I grew up in an abusive family where things were hardly ever what they seemed. "Red and yellow, black and white, they are precious in His sight," the words of one of the many hypocritical Sunday school songs I learned as a child actually meant, "Red is yellow, and black is white, and you can never know for sure what's right." Early on I learned to see incisively into what was not being said and pierce the veil of what was being said. Little did I know how well this would serve me from the year 2000 when I began doing the work I'm doing now.

Derrick Jensen says that the abusive family system is a mirror of the abusive culture and that abusive systems are always set up to protect the abuser. For me, working directly on the wounds of my abuse compelled me to descend very deeply into my own psyche and allow untruths to fall away—not only untruths about my upbringing but also untruths about myself. In the process, I realized that I am comprised of a very necessary human ego as well as a part I call "something greater," that is more than my ego and rational mind.

When my life fell apart at forty, I found the work of Carl Jung and was forever changed by it. Like you, that work allowed me to hold two opposing thoughts in my mind at the same time. I love the way Francis Weller articulates it in his powerful book *Entering the Healing Ground: Grief, Ritual, and the Soul of the World* (80):

> It is indeed the mark of the mature adult to be able to carry these two truths simultaneously. Life is hard, filled with loss and suffering. Life is glorious, amazing, stunning, incomparable. To deny either truth is to live in some fantasy of the ideal or to be crushed by the weight of pain. Instead, both are true, and it requires familiarity with both sorrow and joy to fully encompass the full range of being human.

Many of us register the abuse of the Earth in our own bodies. That may be more evident for people who have suffered abuse—or not. What matters is that abuse is the modus operandi of industrial civilization, and none of us escapes some aspect of it.

For you, natural science, nature, and people represent the trifecta of human experience. For me, the trifecta would be the science of inner exploration, nature, and people. Notwithstanding the fact that I never received a grade higher than a C in any natural science course, I love science and view spirituality and science as a kind of revolving doorway into and beyond our human experience. Matthew Fox says that the purpose of spirituality is to expand and deepen our humanity. But love you as I do, Carl Sagan and Neil de Grasse Tyson, my favorite part of the cosmos is the inner world. Nevertheless, my colleague, Brian Swimme, assists me in savoring both aspects of the universe.

I am fortunate to have people in my life who grasp the reality of near-term extinction and want to live passionately anyway. In addition to my friends, I am part of a group I started five years ago here in Boulder that has evolved from a "collapse preparation" group to an NTE group that we now call Growing Resilience. Like you, I have little contact with my biological family but a great deal of contact with my "adopted" family.

When I visited the Mud Hut, I had the delightful opportunity of meeting some of your friends at the land trust nearby. I was deeply touched by them and their love for you and their passion for the land. Indeed, I believe this is how we were meant to live.

THE NECESSITY OF GRIEVING

More than two decades ago I met African shaman Malidoma Somé. The elders of his Dagara tribe in Burkina Faso in West Africa asked him to move to the US, acquire a Western education, and share the wisdom of his tribe with industrially civilized cultures. Acquire an education he did with two Ph.D.s and two Masters degrees from renowned American universities. The most profound wisdom Malidoma has shared, in my opinion, is the grief ritual regularly practiced in his tribe. The Dagara believe that grieving is essential for the health of the community and that grief not expressed is toxic to the community.

I had the privilege of participating in several Dagara grief rituals and more recently have been asked to facilitate grief rituals in various venues. In fact, doing so has become one of the most rewarding experiences in my current life.

Earlier this year, you participated in a Grief Recovery Training program for several days, and then you and Pauline Schneider conducted a Grief Recovery workshop in Ecuador this spring. I'm wondering how the training was for you and what the workshop in Ecuador was like.

Guy: The Grief Recovery Certification program was spectacular. Its completion allows me to continue serving humanity, albeit in a different capacity than I could have imagined even a few years ago. And the program allowed me to accelerate the processing of my own grief.

I'd long known I was harboring pain, but I could not identify the source. Within the first half day of a four-day session, I realized the source of my pain—grief—and shortly thereafter I had steps to deal with the pain.

Grief is the normal reaction to loss of any kind. In this grief-denying culture, we have developed a large list of myths associated with grief, including that there are two primary sources: death and divorce. But in fact there are more than forty different sources of grief, and I suffered from many of them when I abandoned life as a tenured full professor and commenced life as a homesteader. These many sources of grief result from expectations of more, better, or different outcomes, and they result from incomplete relationships.

In my case, I expected completion of the ongoing collapse of the world's industrial economy before runaway greenhouse was triggered. I expected a better, more vibrant living planet in the wake of civilization's demise. My grief resulted from my unmet expectations. In addition, my grief resulted from unfinished business with respect to relationships. I never allowed myself to say goodbye to the many people who wrote me off when I made the mistake of leaving the job I loved on behalf of the living planet.

Most notably, I lost my identity when I left university life. My position as respected academic was replaced by an unpaid position as digger of ditches, gardener, and general handyman. I wasn't very handy, but I was blessed with a strong back and a weak enough mind to conduct difficult physical work. The disparagement from my family, friends, and colleagues was immediate and profound. I had killed the Buddha—destroyed my own identity—and, like the living planet we're driving to extinction at an accelerating pace, I've yet to recover.

I suffered other losses coincident with losing my ego. I lost essentially all my relationships, my ability to earn money, and the security of a respected, low-work, high-pay job. Although the choice to make that one-way decision was mine, the consequences included a profound sense of personal loss. It persists, but I have learned numerous techniques that facilitate grief recovery, and my regular practice of these techniques greatly aids my own progress.

The foundational document I use to practice and promote grief recovery is *The Grief Recovery Handbook*. The second edition of this book, written by John W. James and Russell Friedman, was published in 2009. It serves as the basis for The Grief Recovery Method®, a version of which I used during the workshop I led in Ecuador in May 2014.

The workshops typically run several weeks, and they follow a format prescribed by James and Friedman, founders of the Grief Recovery Institute®. Among the initial steps are identifying myths about grief and grieving. These include the following: (1) You shouldn't feel bad ("big boys don't cry"); (2) Instead of grieving, replace the loss ("at least you're young enough to have another baby"); (3) You ought to grieve alone; (4) Be strong; (5) Time heals all wounds; and (6) Keep busy. I suspect we're all familiar with these grief-denying strategies that make non-grievers feel better while doing nothing to aid the process of grief recovery.

The latter myth is accompanied by various inappropriate behaviors promoted by contemporary culture. To keep busy, we employ various short-term energy-relieving behaviors (STERBs). Instead of staring into the abyss of our grief, we exercise to excess. Or we drink to excess. Or we lash out in anger. The list goes on, but the point likely is clear: we busily avoid dealing with our grief by shifting the focus to another topic. Any other topic will suffice.

Recovering from grief means achieving completeness in otherwise incomplete relationships. In Grief Recovery Method® workshops, a letter is written to terminate a relationship. The letter is not sent, but it brings closure to a toxic relationship. It may allow a new relationship to develop, assuming the letter is not directed to somebody who has already died.

The workshop in Ecuador was my first, and it was co-facilitated by filmmaker Pauline Schneider. Consistent with my promotion of a gift economy, we conducted the workshop for no charge. We led six participants through an intense, sixteen-hour workshop (with ample homework). Everybody laughed. Everybody cried. And everybody proclaimed success. For an initial offering, I was pleased with the results. I look forward to serving others with future workshops.

I know you are profoundly committed to a life of service, too. What are your next steps?

Carolyn: I'm completely with you on the need for grieving. The hell with storing buckets of beans and barley and amassing weapons. It's really too late for that, but it's never too late for grieving. The masterful Chilean poet, Pablo Neruda, said, "I know the Earth, and I am sad." Neruda wrote this in a moment long before anyone was talking about abrupt climate change, but his words ring exquisitely true today for all of us who are sad because of what is being done to the Earth. In 2011, I visited friends in Cornville, Arizona, and one man in the group shared with us a recording he had just made of bees buzzing in their hive. He had actually lowered a small microphone into the hive and recorded the sound. To my surprise, the very intense buzzing increasingly took on the sound of children crying, and there was an incredibly heartbreaking tone in the buzzing. It pierced by heart and reverberated throughout my body. It was almost unbearable.

Some would say, "Well, you're just projecting your own sorrow onto the noise the bees were making," or "This is the natural sound of bees at work, and there's nothing sad about it." For me, the scientific explanation of what the bees were actually doing is much less relevant than how it affected me and how it triggered my grief. I'd be surprised if the bees weren't expressing their sadness since millions if not billions of them are becoming extinct. But my awareness of the rapidity with which they are going extinct augmented my interpretation of their sound as collective grieving. The bees know the Earth, and perhaps they are sad, too.

People connect with me for life coaching for a variety of reasons, but usually somewhere in the mix is their deep grief. They desperately need safe places to grieve with people who honor and appreciate why they are grieving.

To this end, I am doing more and more grief workshops/ rituals which usually consist of a weekend devoted not only to grieving but to joy and celebration. We begin with a Friday night gathering in which we get to know each other and do some group and individual exercises around grieving. The next morning we do more sharing and group processes, then the actual ritual occurs in the afternoon. It follows the format of the Dagara grief ritual mentioned above and usually concludes in the late afternoon, followed by a time of celebration and sharing a potluck meal. The ritual itself is typically conducted outdoors unless the weather is too cold or snowy. However, it can also happen indoors if necessary. In preparation for the ritual, participants gather the necessary natural materials and work together to create the central structure around which the ritual will center. Overwhelmingly, participants report that they have never experienced anything like it in terms of having permission to grieve and being supported for doing so. A deep sense of community among the group almost always results.

As I also mentioned above, my forthcoming book, *Love in the Age of Ecological Apocalypse: The Relationships We Need to Thrive*, will be published in March 2015. I plan to offer a webinar this fall on the topic, and I'm looking forward to presenting future workshops on the book and facilitating grief rituals in even more venues. Those are two different events, but obviously, they are related.

I love what you said about healing our relationships because as I emphasize in my book, we have a relationship with every person, every object, every activity, every being in our lives, and with the Earth itself. Those relationships are often the source of our grief, even if we are unaware of that reality. I believe that in the time we have left on this planet as near-term extinction bears down upon us, our most important task is healing our relationships. Often we cannot literally heal a breach between ourselves and someone or something else, but we can experience the kind of intra-connectedness that you described above, resulting in a profound sense of personal liberation and gratitude for the relationship. As a result, the heart opens, connections deepen, and we begin to discover meaning and purpose, not only in life, but also in death.

These days, I find that people who have been researching collapse for a few years are almost totally disinterested in doing more research. They are waking up to abrupt climate change and NTE, and I always say that if you haven't had a good existential crisis lately, NTE can certainly provide that. Wherever I go, people tell me that they are weary of collapse research—and even climate research—and are seeking support for living from a hospice perspective.

In the light of NTE, I am thinking a great deal these days about what NTE means for us as elders. As you probably have noticed, I've written in many places about elderhood and have distinguished it from age. In other words, "older" doesn't necessarily mean "elder." Having embraced the more indigenous definition of elderhood, I have encountered many people in their twenties and thirties whom I consider elders because, for me, elderhood has little to do with age and much more to do with wisdom. We all know people in their seventies and eighties who on many levels have never grown up, and at the same time we have met much younger people who are wise beyond their years.

I'm wondering how you view elderhood and the role of being an elder. What responsibilities lie before us as elders? Have you also met "youngers" who are in many ways "elders"?

Guy: I've met many people who are simultaneously younger than me and wiser than me. One of the reasons I pursued anarchy in my classrooms was to point out that we each have worthy contributions. Over the years, I came to understand and treat each group of people with whom I was fortunate to work as a corps of discovery. Our quest? A life of excellence for each of us.

Naturally, most students did not welcome the quest, at least not initially. In fact, I met considerable resistance each semester as I explained how a life of excellence can and should be pursued along the path of Fire Management or Conservation Biology or Sustainable Living (to list a few titular examples of courses I led recently). But by the end of the second week of each semester, resistance imposed by culture was overcome by the joy, humor, and richness of a liberal approach that valued contributions from every participant during every meeting.

An example might help. In February 2006, during the ninth meeting of a class titled Wildland Vegetation Management, the syllabus indicated the day's topic was conspicuous consumption. Already we're on tenuous ground from the perspective of the typical university administrator. During the first few minutes that day, somebody mentioned Siddartha Gautama (i.e., the Buddha) and his four noble truths. The link to conspicuous consumption should be apparent. Less apparent are the following topics, all of which we addressed within the first ten minutes of the class period, in this order:

- No Child Left Behind (federal legislation)
- Culverts under Speedway (a main surface street in Tucson)
- *The Princess Bride* (film)
- Charlton Heston (actor)
- Sheryl Crow (singer)
- John Dewey (pragmatist philosopher and educator)
- Espresso Art (local coffee shop)
- New Jersey Turnpike
- Socrates
- Pangaea

The conversation flowed naturally from one topic to another, and all topics were linked directly to the idea of conspicuous consumption. As traditionally taught, Wildland Vegetation Management is focused upon developing the means to continue our conspicuous consumption. As a group, we were questioning the validity of this historical approach within the context of popular culture and our own experiences.

As you might expect, I was not allowed to teach the course again (as is the department's head prerogative). Indeed, within a few months after the course was completed, and shortly after we hired a new department head, she banned me from teaching any classes in the department. Despite rhetoric, the contrary, the "sage on the stage" model of teaching is still preferred in many institutions of "education."

This minor example illustrates an important point: We learn when we listen, not when we talk. And each of us has a lot to learn.

Elders have much to offer because they have accumulated life experiences. They have even more to offer if they have pondered those life experiences and are able to place them into historical and cultural contexts. Stories about the "good old days" have abundant meaning when compared to contemporary circumstances.

In tribal communities, elders are revered for two reasons. First, elders have contributed to the tribe with their younger, more-able bodies. Second, they continue to contribute to the tribe with their minds. In my own community, able-bodied youngsters respect and willingly assist their elders (perhaps in part because the younger set is familiar with the notion of "paying it forward").

We elders are charged with many duties. Demonstrating humility is among them, but demonstrating hubris is not. After all, our generation leaves in our wake a world depleted of finite materials, ruined by empire, and ruled by fascism masquerading as republic. We can guide, but we cannot push. We can offer, but we cannot force somebody to accept the offer. And considering our record, I think our role as guides is suspect at best.

Carolyn: Starting in about 2006, I began all of my psychology and history classes by apologizing to my students for the world my generation had left them. Many did not understand or even have a clue why I was making such an apology, let alone why I was beginning the class with it. By the end of the semester when I reminded them of the first day of class, they understood the reason for the apology, and many thanked me.

Sadly, there are few places on the planet where a young person can still experience a formal rite of passage. How can young people be initiated when there are no initiated elders to facilitate and guide the experience? I believe that ninety of what we hate about industrial civilization could have been averted had even a small percentage of the traditional cultures of the world remained intact.

After traveling around the world and observing these cultures, Jung stated that he believed something in us wants and needs initiation, and that when we aren't able to experience a formal initiation, we are "initiated" by life in terms of major life losses such as divorce, grave illness, the loss of loved ones—and in the twenty-first century experiences such as loss of employment, bankruptcy, foreclosure, and of course the loss of ecosystems. Birth and death are also initiatory experiences. With each initiation, we have an opportunity to "mine the gold" in the loss for the purpose of becoming more conscious, compassionate, and Earth-centered human beings.

Ecopsychologist and wilderness guide Bill Plotkin states in *Spiritual Ecology: A Cry for the Earth*, "An authentic adult is someone who experiences herself, first and foremost, as a member of the Earth community." Again, this recognition has little to do with age, and much more to do with wisdom. Plotkin continues, noting the following:

> *Those who have attained true elderhood are the ones most gifted at understanding the needs and desires of the Earth. True elders can hear the Earth whispering her guidance. Especially at this time of great crisis and transition, it is the Earth's guidance that we most need (194)....Elders know that personal maturation often begins with a dying, a positive disintegration, always a necessity before a rebirth...The elders of our time must assume the role of underworld guides for the human collective in our current planetary rite of passage. (198-199)*

For several years, I have been writing and speaking about our current collapse and extinction as forms of planetary initiation. In no way does this mean that these crises will result in a mass transformation of consciousness. I believe that some people will experience a spiritual rebirth, but only as a result of death—whether that is the death of their "civilized" perspective or a literal death. It is important to notice that in traditional cultures, initiation was always dangerous because it always involved some risk. Not everyone survived it physically, but the elders of those cultures believed that the rebirth it could produce in its youth was worth the risk.

My job, and I suspect yours as well, is to help people die well—whether that involves the death of the civilized ego or an actual termination of physical life. At the same time, it seems that both of us are also committed to supporting people in living well, which for me means living with meaning and purpose with a commitment to compassionate service to the human and more-than-human worlds. It also means creating joy and beauty wherever and whenever possible.

In fact, we should probably talk about what living well means from both our perspectives.

LIVING AND DYING WELL

Guy: To start, I recommend allowing yourself to feel the emptiness wash over you. Get off your smart phone. Put aside all your digital toys and immerse yourself in the horrors of being alone. We're alone. It's tragic. The universe doesn't care about us as a species, much less about us as individuals.

Feel the pain of insignificance. And then perhaps you can appreciate the joys of being here at all, against overwhelming odds.

How overwhelming are the odds against our existence? We understand DNA well enough to know that the odds against any of us appearing in physical form exceed the odds against plucking a single atom, at random, from the entire universe. If I believed in miracles, I'd have to think each of us represents a miracle. To quote evolutionary biologist Richard Dawkins, "In the teeth of these stupefying odds it is you and I that are privileged to be here, privileged with eyes to see where we are and brains to wonder why."

If a line from learned professor Richard Dawkins isn't sufficiently convincing, perhaps one from boxer Mike Tyson will do the trick: "Everybody has a plan until they get punched in the face."

Industrial civilization punched the planet in the face, repeatedly. We wonder why Earth isn't the same planet with which we grew up. We offer apologies, including recycling and hybrid automobiles, but we're unwilling—as a society—to pursue substantive change. We'll apologize and negotiate, but we'll not make sacrifices to terminate civilization, at least not at scale.

And now it's too late. Earth is done with humans. We're walking around to save money on funeral expenses.

My response? Pursue radicalism. Get to the root. Pursue a counter-cultural path. Pursue excellence, in the spirit of Socrates.

Pursue what you love. Throw off the cultural shackles. All of 'em.

We're all in hospice now. Let's give freely of our time, wisdom, and material possessions. Let's throw ourselves into humanity and the living planet. Let's act with compassion and courage. Let's endow ourselves with dignity. Even if all the data, models, assessments, and forecasts about abrupt climate change are incorrect, even if Earth can support infinite growth on a finite planet with no adverse consequences, I remain unconvinced there is a better way to live.

And let's not forget the immortal words of writer Edward Abbey: "Action is the antidote to despair." After all, I am not now suggesting, nor have I ever suggested, giving up. Our insignificant lives have never been about us. They're about the shards of the living planet we leave in our wake. As pointed out by Desmond Tutu, "If you are neutral in situations of injustice, you have chosen the side of the oppressor."

Anybody can cheer for winners. Essentially everybody I know sits on the sidelines cheering for the favorites. I'm proposing we stop sitting on the sidelines. I'm proposing we get to work on behalf of the underdogs. I'm proposing we start working to save the living planet.

Is there a better metric of a person's character than how she treats those who can do nothing for her? Let's treat others with respect as we leave this mortal coil. Let's extend the concept of "other" to humans within non-industrial cultures. Let's extend the concept further, to non-human species. You might think they've been doing nothing for us, but they've been providing for our own existence. Let's return the favor.

In the end, for finite beings such as ourselves, the historical process is irrelevant; all we have is our legacy, but that legacy is lost to us (as individuals). Yet we are unique beings in that we are able to recognize the historical process as something larger than ourselves. We judge that process worthy or not worthy based on our own singular experience. For me, the universe is a worthy endeavor because the lens through which I view it is colored with the relationships I have experienced. Those relationships include humans and nature.

Consider how I give away my time. I deliver presentations in exchange for travel expenses. I sleep in the homes of my hosts, and eat at their tables. This minimizes monetary expenditures and also allows us to interact in a meaningful manner.

I've slept in nice hotels, superb B&Bs, and wonderful homes. I've also slept on a couch that seemingly harbored at least a dozen sexually transmitted diseases.

I've dined at the tables of many hosts, and enjoyed superb meals lovingly crafted. I've dined in restaurants that should've been closed by the state health department. I've gained fifteen pounds on a single, two-week speaking tour. And I've gone thirty hours without eating because my hosts weren't paying attention.

Most importantly, I've met amazing people and participated in deeply meaningful conversations. Sharing experiences, sharing meals, and sharing ideas constitute the best of humanity and the best memories of my life. The people I've met along the way enrich my life beyond measure.

Accepting our mortality as a gift is hardly a new idea. Social critic Jonathan Swift poignantly described the horrors of immortality in his 1726 book, *Gulliver's Travels*. Many other writers have subsequently joined the fray. Contemporary American poet Mary Oliver points out, "Someone I loved once gave me a box full of darkness. It took me years to understand that this too, was a gift."

Our own death is a gift for several reasons, most notably because it means we get to live. Let's live. In the spirit of seizing the moment, let's live now. In the spirit of awareness and in recognition of this moment, let's live here now.

Carolyn: Thank you for quoting my favorite American poet, Mary Oliver. As you and probably everyone reading this book knows, I use poetry a great deal in my work because it so effectively and effortlessly transports us out of the left brain and into our hearts. Let me hasten to add that for me, there's nothing wrong with using the left brain with which your contributions to this book are replete. However, I'm noticing that you're also adding much more from the heart than you might have before the Grief Recovery training and before your encounter with your wonderful Buddhist friends in Winnipeg.

In this book, you are presenting us with climate science realities that require constant updating because they change almost daily. Conversely, I'm offering here timeless, ancient wisdom that needs no updating but only heartful application to our current predicament. Industrial civilization is relentlessly seeking "new ideas" and "innovations" which will improve itself. Indigenous traditions are forever turning to perennial wisdom inferred for the most part from their intimate relationship with the Earth.

Someone recently sent me a link to the trailer of a new documentary called *The Making of Humans*, with Stephen Jenkinson and produced by filmmaker Ian Mackenzie. A teacher, author, and storyteller, Jenkinson has also worked extensively in palliative care and was featured in 2012 in another short film entitled *Griefwalker*, which was filmed over a twelve-year period and takes the viewer into his exchanges with dying patients. Along the way, Jenkinson suggests that grief is more than a feeling. It's a skill, and I believe that conscious grieving is a skill we must learn and practice purposefully in the face of near-term extinction.

Grounded in a hospice perspective, Jenkinson says in *The Making of Humans* that "To see what you love dying and to continue to love it when it's not going to last, that in itself is an act of love. That could be part of the story."

For me, these two sentences encapsulate what we have been articulating in this book. We see what is dying, and we love it anyway, and not just "anyway," but because it is dying. When you know you are losing a human being you love, your life becomes all about loving that person with all your heart, soul, mind, body, and breath.

One of the most important ways we can love our dying planet and our dying species is to live passionately in this time of extinction. As we have stated above, that means living a life of service, creating joy and beauty at every possible opportunity, and unleashing our creativity in spite of our culturally programmed messages about it not being "good enough." And one of the most profound ways to live passionately is to fall deeply in love with the Earth—for the first time or the second time or as often as we can.

As a result, we will, as you have said many times, work to make this demise easier on other species. In his wonderful essay, "In Praise of Manners," Francis Weller states the following:

> *If we are leaving, then we owe it, as an act of respect and manners, to try to do whatever we can to mitigate against further damage. If our species is leaving, there will be others that remain. And if the salmon are returning, we should do whatever we can to make their waters clean and take down the obstacles that interfere with their ability to spawn. We should do whatever we can to make sure the forest can go back to being a full climax forest where all species are able to return. We should do whatever we can and I think that is an act of deep manners. And again we come back to the heart, don't we?*

For me, the absolute bottom line is what I believe we came here to do. Even beyond the service, joy, beauty, and creativity we can offer, we are fundamentally here to experience death and rebirth—the death of who we think we are so that who we are in our core, that is our sacred Self, may be born. Even if there were no catastrophic climate change or near-term extinction breathing down our necks, that is, I believe, our mission on this planet.

END

BIBLIOGRAPHY

Amberg, Ann. What Does The Universe Do? https://
sites.google.com/site/annambergdesign/

Baker, Carolyn. "Embarking On The Journey Of Consciousness:
Staying On The Train." Speaking Truth To Power. January
13, 2014. http://www.carolynbaker.net/2014/01/13/
embarking-on-the-journey-of-consciousness-staying-on-the-
traing/

Baker, Carolyn. Love In The Age Of Ecological Apocalypse: The
Relationships We Need To Thrive. Berkeley, California:
North Atlantic Books, 2015.

Baker, Carolyn. "Mad Hominem: Why Hatred Of The Human
Species Is Not Helpful." Speaking Truth To Power. March 9,
2014. http://www.carolynbaker.net/2014/03/09/mad-
hominem-why-hatred-of-the-human-species-is-not-helpful-
by-carolyn-baker/

Baker, Carolyn. "Maintaining Mental Health In The Age Of
Madness." Speaking Truth To Power. March 18, 2013. http://
www.carolynbaker.net/2013/03/18/maintaining-mental-
health-in-the-age-of-madness-by-carolyn-baker/

Baker, Carolyn. "What Does It Mean To Do Something About
Climate Change?". Speaking Truth To Power. April 13,
2014. http://www.carolynbaker.net/2014/04/13/what-does-it-
mean-to-do-something-about-climate-change-by-carolyn-
baker/

Barrett, Scott, et. al, "Climate Engineering Reconsidered," Nature Climate Change (online), 25 June 2014

Bolden, Charles, F., "It's Vital We Become a "multi-planet species," NASA/Time Video, April 24, 2014

Brysse, Keynyn, et. al, "Climate change prediction: Erring on the side of least drama?" Global Environmental Change, Volume 23, Issue 1, February 2013, Pages 327–337

Buzzell, Linda and Chalquist, Craig. Eco-Therapy: Healing With Nature In Mind. Berkeley, California: Counterpoint Press, 2009.

Choi, Charles, Q., "Geoengineering Ineffective Against Climate Change, Could Make Worse," Livescience, February, 25, 2014

Clement, Scott, "How Americans see global warming — in 8 charts," The Washington Post, April 22, 2013

Dilingpole, James, "An English class for trolls, professional offence-takers and climate activists," The Telegraph (online), April 7th, 2013

Edwards, Bobbie, "Geoengineering Would Cool Planet, But Screw Up Rainfall Patterns, MongaBay.com, January, 14, 2014

Eisenstein, Charles. Sacred Economics: Money, Gift, and Society In the Age of Transition. Berkeley, California: Evolver Editions of North Atlantic Books, 2011.

Eisenstein, Charles. The Ascent Of Humanity. Harrisburg, Pennsylvania: Panenthea Press, 2007.

Feffer, John, "Earth: Game Over?" Truthout, 27 April 2014

Garrett, Tim, "Is Global Warming Unstoppable? Theory Also Says Energy Conservation Doesn't Help," University of Utah, 2009

Goethe, Johann Wolfgang von, Die Wahlverwandtschaften (Elective Affinities) Bk. II, J. G. Cottaische Buchhandlung, Berlin1809

Goldblatt, Colin, et. al, "Low simulated radiation limit for runaway greenhouse climates," Nature Geoscience, 28 July 2013

Goldenberg, Suzanne, "Arctic lost record snow and ice last year as data shows changing climate findings from US science agency Noaa suggest widespread and irreversible changes because of a warming climate," The Guardian, 5 December 2012

Goldenberg, Suzanne, "UN: rate of emissions growth nearly doubled in first decade of 21st centuryLeaked draft shows emissions grew nearly twice as fast from 2000-10 as in previous 30 years – despite economic slowdown," The Guardian, 11 April 2014

Harball, Elizabeth, "Could Carbon Farms Reverse Global Warming? Large plantations might pull CO2 out of the air," Scientific American, August 23, 2013

Hamilton, Clive Dr., "Earthmasters: The Dawn of the Age of Climate Engineering," Yale University Press (April 22, 2013)

Hedges, Chris, "American Socrates," Truthdig, June 15, 2014

Holocek, Andrew. Preparing To Die: Practical Advice And Spiritual Wisdom From The Tibetan Buddhist Tradition. Boston, Massachussetts: Shambhala Publications, 2013

Hunziker, Robert, "Inevitable Surprises: Looming Danger of Abrupt Climate Change," Counterpunch, December 26, 2013

James, John and Friedman, Russell. The Grief Recovery Handbook. New York: Harper Collins, 2009.

Jennings, Michael Dr., "Climate Dark Age Summary," Echoshock Radio, March 19, 2014

Jennings, Robert, "Climate Disruption: Are We Beyond the Worst Case Scenario?" Global Policy, 3 September 2012

Jensen, Robert. "After The Harvest: Learning To Leave The Planet Gracefully." Common Dreams, June 14, 2014. http://www.commondreams.org/views/2014/06/14/after-harvest-learning-leave-planet-gracefully

Khan, Shfaqat, "Sustained mass loss of the northeast Greenland ice sheet triggered by regional warming," Nature Climate Change, 16 March 2014

Kingsley, Peter. Ancient Philosophy, Mystery And Magic: Empedocles And The Pythagorean Tradition. Oxford: Oxford University Press, 1997.

Kingsley, Peter. A Story Waiting To Pierce You. Golden Sufi Center, 2010

Kingsley, Peter. In The Dark Places Of Wisdom. Golden Sufi Center, 1999.

Kingsley, Peter. Reality. Point Reyes Station: Golden Sufi Center, 2004.

Kolbert, Elizabeth, "Is It Too Late to Prepare for Climate Change?" The New Yorker, November 4, 2013

Leakey, Louise, "WATCH: Is The Human Race In Danger Of Becoming Extinct Soon?" Huffington Post, July 2013

Lee, Howard, "Rapid Climate Changes More Deadly Than Asteroid Impacts in Earth's Past—Study Shows," Skeptical Science, 27 May 2014

Lee, Howard, "Alarming New Study Makes Today's Climate Change More to Earth's Worst mass Extinction," Skeptical Science, 2 April 2014

LePage, Michael, " "How much hotter is the planet going to get?" New Scientist, 9 March 2014

Levy, Hiram, "The roles of aerosol direct and indirect effects in past and future climate change," Journal of Geophysical Research, 20 May 2013

Levy, Paul. Dispelling Wetiko: Breaking The Curse Of Evil. Berkeley, California: North Atlantic Books, 2013.

Light, Malcolm, "The Non-Disclosed Extreme Arctic Methane Threat," Runaway Global Warming, 22 December 2013

Livinal, V.N., Lenton, T.M., "A recent tipping point in the Arctic sea-ice cover: abrupt and persistent increase in the seasonal cycle since 2007," The Cryosphere 2013

McPherson, Guy. "Only Love Remains." Nature Bats Last. October 4, 2012. http://guymcpherson.com/2012/10/only-love-remains/

McPherson, Guy. Walking Away From Empire: A Personal Journey. Frederick, Maryland: America Star Books, 2011.

Melton, Bruce, "Abrupt Climate Change: No Bioperturbation," Truthout, 18 March 2014

Melton, Bruce, "Where We Are Now - Not What You Think," Climate Change, 26 December 2013

Mooney, Chris. "The Science Of Why We Don't Believe Science." Mother Jones. May/June, 2011. http://www.motherjones.com/politics/2011/03/denial-science-chris-mooney

Morales, Alex, "Kyoto Veterans Say Global Warming Goal Slipping Away," Bloomberg, November 4, 2013

Morlighem, M., "Deeply incised submarine glacial valleys beneath the Greenland ice sheet," Nature Geoscience, 18 May 2014

National Oceanic and Atmospheric Administration (NOAAA), "New Study Shows Climate Change Largely Irreversible," NOAAA, January 26, 2009

O'Donohue, John. Anam Cara: A Book Of Celtic Wisdom. San Francisco, Harper Collins, 1997.

Plotkin, Bill. Nature And The Human Soul: Cultivating Wholeness And Community In A Fragmented World. Novato, California: New World Library, 2007.

Pollard, Dave. "In Defense Of Inaction: How To Save The World." April 20, 2014. http://howtosavetheworld.ca/2014/04/20/in-defence-of-inaction/

Praetorius, Summer K. et al, "Synchronization of North Pacific and Greenland climates preceded abrupt deglacial warming," Science, 25 July 2014

Quintero, Ignacio, Wiens, John J., "Rates of projected climate change dramatically exceed past rates of climatic niche evolution among vertebrate species," Ecology Letters, 26 June 2013

Romm, Joseph, "Hadley Center study warns of 'catastrophic' 5-7°C warming by 2100 on current emissions path," Grist 23 December 2008

Roy, Arundhati, *Power Politics* (Second Edition), South End Press; Second Edition edition (April 1, 2002)

Schwartz, Larry, "10 of the biggest threats to human existence, just in case you don't have enough to worry about," Alternet, July 26, 2014

Scribbler, Robert, "Global Sea Surface Temperatures Increase to Extraordinary +1.25 C Anomaly as El Nino Tightens Grip on Pacific," Climate Change, May 22, 2014

Spratt, David, "Is climate change already dangerous?" Climate Code Red, September 2013

Stanford University, ""Climate change occurring ten times faster than at any time in past 65 million years," ScienceDaily, 1 August 2013

Steiner, Susie. "Top Five Regrets Of The Dying." The Guardian, February 1, 2012. http://www.theguardian.com/lifeandstyle/2012/feb/01/top-five-regrets-of-the-dying

Stephenson, Wen. "Let This Be the Last Earth Day." Nation Magazine, April 22, 2014. http://www.thenation.com/blogs/wen-stephenson

Summerhayes, Colin, "Earth's sensitivity to climate change could be 'double' previous estimates, say geologists," The Geological Society, 10 December 2013

The Powers Of The Universe. Brian Swimme. Center For The Story Of The Universe, 2004.

Van Groenigen, Kees Jan, "Faster Decomposition Under Increased Atmospheric CO2 Limits Soil Carbon Storage," Science, 2 May 2014

University of Cambridge, "Global warming: Four degree rise will end vegetation 'carbon sink', research suggests," ScienceDaily, 16 December 2013

University of Toronto, ""New long-lived greenhouse gas discovered: Highest global-warming impact of any compound to date," Science Daily, 9 December 2013

Vaughan-Lee, Llewellyn. Spiritual Ecology: The Cry Of The Earth. Point Reyes Station, California: Golden Sufi Center, 2013.

"Welcome To The Anthropocene." Owen Gaffney. n.d http://www.anthropocene.info/en/anthropocene

Weller, Francis. Entering The Healing Ground: Grief, Ritual, And The Soul Of The World. Santa Rosa, California: Wisdom Bridge Press, 2011.

Woodbury, Zhiwa. "Planetary Hospice: Rebirthing Planet Earth." Ecological Buddhism. March, 2013. http:// www.ecobuddhism.org/science/featured_articles1/ph1

World Bank, "Turn down the heat : why a 4°C warmer world must be avoided," World Bank, 2012

BIOGRAPHIES

Carolyn Baker, Ph.D. is a former psychotherapist and professor of psychology and history. She is the author of *Collapsing Consciously: Transformative Truths For Turbulent Times, Navigating The Coming Chaos: A Handbook For Inner Transition* (2011) and *Sacred Demise: Walking The Spiritual Path Of Industrial Civilization's Collapse* (2009). Carolyn offers life coaching for people who want to live more resiliently in the present as they prepare for the future.

Guy McPherson is Professor Emeritus at the University of Arizona. He taught and conducted research for twenty award-winning years before leaving the university for ethical reasons in 2009. McPherson established a homestead and continues his prolific writing and teaching from there.

NEXT REVELATION PRESS

EXTINCTION DIALOGS

HOW TO LIVE WITH DEATH IN MIND

CAROLYN BAKER & GUY McPHERSON

CPSIA information can be obtained at www.ICGtesting.com
Printed in the USA
LVOW05s0239131214

418616LV00019B/131/P

paint
particulate mask